Case-based Study for
Radiologist and
Cardiologist

Case-based Study for
Radiologist and
Cardiologist

Reena Anand

DNB, DMRD, MNAMS

Principal Consultant, Cardiovascular Imaging
Department of Radiology
Max Super Specialty Hospital
Saket, New Delhi, India

CBS

CBS Publishers & Distributors Pvt Ltd

New Delhi • Bengaluru • Chennai • Kochi • Kolkata • Mumbai
Bhubaneswar • Hyderabad • Jharkhand • Nagpur • Patna • Pune • Uttarakhand

Disclaimer

Science and technology are constantly changing fields. New research and experience broaden the scope of information and knowledge. The author has tried her best in giving information available to her while preparing the material for this book. Although all efforts have been made to ensure optimum accuracy of the material, yet it is quite possible some errors might have been left uncorrected. The publisher, the printer and the author will not be held responsible for any inadvertent errors or inaccuracies.

Case-based Study for
Radiologist and
Cardiologist

ISBN: 978-93-87742-91-8

Copyright © Author and Publisher

First Edition: 2018

Published by Satish Kumar Jain and Produced by Varun Jain for

CBS Publishers & Distributors Pvt Ltd
4819/XI Prahlad Street, 24 Ansari Road, Daryaganj, New Delhi 110 002, India.
Ph: 23289259, 23266861, 23266867   Fax: 011-23243014   Website: www.cbspd.com
e-mail: delhi@cbspd.com; cbspubs@airtelmail.in.
Corporate Office: 204 FIE, Industrial Area, Patparganj, Delhi 110 092
Ph: 4934 4934   Fax: 4934 4935   e-mail: publishing@cbspd.com; publicity@cbspd.com

Branches

• **Bengaluru:** Seema House 2975, 17th Cross, K.R. Road,
 Banasankari 2nd Stage, Bengaluru 560 070, Karnataka
 Ph: +91-80-26771678/79   Fax: +91-80-26771680   e-mail: bangalore@cbspd.com
• **Chennai:** 7, Subbaraya Street, Shenoy Nagar, Chennai 600 030, Tamil Nadu
 Ph: +91-44-26680620, 26681266   Fax: +91-44-42032115   e-mail: chennai@cbspd.com
• **Kochi:** Ashana House, No. 39/1904, AM Thomas Road, Valanjambalam,
 Ernakulam 682 016, Kochi, Kerala
 Ph: +91-484-4059061-65   Fax: +91-484-4059065   e-mail: kochi@cbspd.com
• **Kolkata:** 6/B, Ground Floor, Rameswar Shaw Road, Kolkata-700 014, West Bengal
 Ph: +91-33-22891126, 22891127, 22891128   e-mail: kolkata@cbspd.com
• **Mumbai:** 83-C, Dr E Moses Road, Worli, Mumbai-400018, Maharashtra
 Ph: +91-22-24902340/41   Fax: +91-22-24902342   e-mail: mumbai@cbspd.com

Representatives

| • **Bhubaneswar** | 0-9911037372 | • **Hyderabad** | 0-9885175004 | • **Jharkhand** | 0-9811541605 | • **Nagpur** | 0-9021734563 |
| • **Patna** | 0-9334159340 | • **Pune** | 0-9623451994 | • **Uttarakhand** | 0-9716462459 | | |

Printed at: Magic International Pvt. Ltd., Greater Noida, UP, India

Contributors

Bharat Aggarwal
Director, Department of Radiology
Max Super Specialty Hospital, Saket
New Delhi, India

KK Talwar
Chairman, Department of Cardiology
Max Super Specialty Hospital, Saket
New Delhi, India

Rajneesh Malhotra
Senior Director, Department of Cardiothoracic
Surgery
Max Super Specialty Hospital, Saket
New Delhi, India

Kumud Rai
Senior Director, Department of Vascular Surgery
Max Super Specialty Hospital, Saket
New Delhi, India

KS Dagar
Director, Department of Paediatric Cardiac Surgery
Max Super Specialty Hospital, Saket
New Delhi, India

Viveka Kumar
Senior Director, Department of Cardiology
Max Super Specialty Hospital, Saket
New Delhi, India

Kewal Krishan
Director, VADs and Heart Transplant
Department of Cardiothoracic Surgery
Max Super Specialty Hospital, Saket
New Delhi, India

CP Roy
Director, Department of Cardiology
Max Super Specialty Hospital, Saket
New Delhi, India

Sonia Dhall
Head, Department of Radiology
Max Super Specialty Hospital, Saket
New Delhi, India

Rajesh Chand
Director, Department of Cardiac Anaesthesia
Max Super Specialty Hospital, Saket
New Delhi, India

Rajeev Rathi
Director, Department of Cardiology
Max Super Specialty Hospital, Saket
New Delhi, India

Amit Kumar Sahu
Consultant, Department of Radiology
Max Super Specialty Hospital, Saket
New Delhi, India

Roopa Salwan
Senior Director, Department of Cardiology
Max Super Specialty Hospital, Saket
New Delhi, India

Sumeet Sethi
Senior Consultant, Department of Cardiology
Max Super Specialty Hospital, Saket
New Delhi, India

Vanita Arora
Director, Department of Cardiology
Max Super Specialty Hospital, Saket
New Delhi, India

Neeraj Avasthi
Principal Consultant, Department of Paediatric
Cardiology
Max Super Specialty Hospital, Saket
New Delhi, India

Mohan Bhargava
Senior Consultant, Department of Cardiology
Max Super Specialty Hospital, Saket
New Delhi, India

Sunil Aggarwal
Senior Consultant, Department of Cardiology
Max Super Specialty Hospital, Saket
New Delhi, India

Vivek Saxena
Head, Intervention Radiology
Max Super Specialty Hospital, Saket
New Delhi, India

Anandmoyee Dhar
Principal Consultant, Department of Radiology
Max Super Specialty Hospital, Saket
New Delhi, India

Richa Bansal
Principal Consultant, Department of Radiology
Max Super Specialty Hospital, Saket
New Delhi, India

Amit Kumar
Principal Consultant, Department of Radiology
Max Super Specialty Hospital, Saket
New Delhi, India

Abhishek Aggarwal
Senior Consultant, Department of Radiology
Max Super Specialty Hospital, Saket
New Delhi, India

Gurpreet Makkar
Principal Consultant, Department of Radiology
Max Smart Hospital, Saket
New Delhi, India

Ruchi Rastogi
Principal Consultant, Department of Radiology
Max Super Specialty Hospital, Saket
New Delhi, India

Ganesh Mani
Chairman, Department of Cardiothoracic Surgery
Max Smart Hospital, Saket
New Delhi, India

SK Sinha
Director, Department of Cardiothoracic Surgery
Max Smart Hospital, Saket
New Delhi, India

Ripen Gupta
Senior Consultant, Department of Cardiology
Max Super Specialty Hospital, Saket
New Delhi, India

Vivek Tandon
Director Echocardiography
Department of Cardiology
Max Smart Hospital, Saket
New Delhi, India

Anupam Goyal
Senior Consultant, Department of Cardiology
Max Super Specialty Hospital, Saket
New Delhi, India

Sanjeev Malhotra
Senior Consultant
Department of Cardiothoracic Surgery
Max Super Specialty Hospital, Saket
New Delhi, India

Bharat Bansal
Attending Consultant
Department of Non-invasive Cardiology
Max Super Specialty Hospital, Saket
New Delhi, India

Ratna Malika Kumar
Senior Consultant
Department of Cardiothoracic Surgery
Max Super Specialty Hospital, Saket
New Delhi, India

Rahul Mehrotra
Principal Consultant and Head
Department of Non-invasive Cardiology
Max Super Specialty Hospital, Saket
New Delhi, India

Raj Kumar
Consultant, Echocardiography Lab
Department of Cardiology
Max Super Specialty Hospital, Saket
New Delhi, India

Kapil Gupta
Consultant, Department of Vascular Surgery
Max Super Specialty Hospital, Saket
New Delhi, India

Himanshu
Consultant, Department of Paediatric Cardiology
Max Super Specialty Hospital, Saket
New Delhi, India

Ritwique Bhuyan
Principal Consultant
Department of Cardiothoracic Surgery
Max Hospital, Patparganj, New Delhi, India

Naresh Goel
Senior Consultant, Department of Cardiology
Max Hospital, Shalimar Bagh
New Delhi, India

Rajesh Gothi
Senior Consultant, Department of Radiology
Holy Family Hospital, New Delhi, India

Aparna Juneja
Associate Consultant, Department of Radiology
Asian Hospital, NCR, India

Pawan Kumar Singh
Senior Resident
Department of Cardiothoracic Surgery
Max Super Specialty Hospital, Saket
New Delhi, India

Showkat Bhat
Attending Consultant
Department of Non-invasive Cardiology
Max Super Specialty Hospital, Saket
New Delhi, India

Sachin Chatterjee
Attending Consultant
Department of Cardiothoracic Surgery
Max Super Specialty Hospital, Saket
New Delhi, India

Foreword

It is a pleasure to write the Foreword for *Case-based Study for Radiologist and Cardiologist* in which we intend to review cardiac imaging from the perspective of the various imaging modalities usually performed to obtain anatomic and functional information of the heart.

This book is designed to provide truly useful information to radiologists, cardiologists, residents and fellows. This book represents a new and different way of learning cardiovascular imaging. Topics in the book include myocardial diseases, cardiac masses, clots, and pericardial diseases, ischemic and non-ischemic cardiomyopathies. Demonstrating the values and limitations of the imaging techniques, the book enables practitioners to determine which test, in which patient population, and for which purpose would be the most appropriate to use.

Case-based Study for Radiologist and Cardiologist is an excellent guide and source for residents, fellows, and practitioners in both radiology and cardiology and for those who want to refresh their knowledge in the multimodality field of cardiovascular imaging.

Dr Bharat Aggarwal
Director, Max Super Specialty Hospital, Saket
New Delhi, India

Preface

The rapid advance in both MR and CT technology and the constantly evolving concepts on their clinical use has made it mandatory to keep updating knowledge of cardiovascular imaging.

Our goal in preparing this case-based book is to provide updated information on the use of MR and CT for the assessment of cardiac and vascular diseases.

This case-based book spans a series of cases from their initial presentation to management.

The authors include radiologist, cardiac surgeons and cardiologists having a long involvement and interest in fostering the development of cardiovascular MR and CT.

Reena Anand

Acknowledgements

Acknowledgements are very essential because all those who have made contributions, given encouragement, inspiration, guidance and support must be duly recognized, thanked and remembered. It will be a sort of ingratitude on my part if I ignore those who have constantly inspired me and thus helped me develop the idea which was in embryo in my mind and is now in full and blooming shape before you. At once two names crop up in mind.

Dr Bharat Aggarwal (Director, Radiology, Pan Max Hospitals) who infused this brilliant idea and guided me to compile the presentation of cardiac cases made by me in the Max Hospital in the span of last three years from 2014 to 2017 in the form of a book. He advised me to carve the presentations into chapters so that there is no overlapping. Without his inspiration, I would have rambled in the fairy land and never touched the solid ground.

It will be my shortsightedness and ingratitude if I do not give recognition and eulogy to the chairman Dr KK Talwar (Cardiology, Pan Max Hospitals) for his support and guidance. He gave the suggestion of conducting clinical cardioradiology meeting every month where the cardiac case were discussed and all these cases I have compiled into this book.

I am also grateful to my family especially my husband Dr Kewal Krishan, Director, LVAD and Heart Transplant, Pan Max Hospitals who time and again hinted, goaded and made me fully prepared to rise up with full zeal and vigor to put my thoughts into real shape.

I offer my sincere thanks to all the colleagues particularly Dr Sonia Dhall, Head, Department of Radiology, Max Hospital, Saket, for data and reference and Dr Amit Sahu, Consultant Radiology, Max Hospital, for his help in preparation of this manuscript.

I am heartily thankful to CBS Publishers & Distributors, especially we would like to put on record the sincere efforts of Shree Satish Kumar Jain (Matta ji), CMD and Shree Varun Jain (Director), CBS Publishers & Distributors, New Delhi, for their patience, goodwill and cooperation. I express my gratitude to Mr YN Arjuna (Senior Vice President Publishing, Editorial and Publicity); Mrs Ritu Chawla (AGM Production); Mr Sunil Dutt; Mr Parmod Kumar and Mr Rohan Prasad, for their skillful service and immense help in editing and figure work of the manuscript.

Reena Anand

Contents

Contributors   *v*

Foreword by Dr Bharat Aggarwal   *vii*

Preface   *ix*

List of Abbreviations   *xv*

Protocols   *xvii*

1. Arrhythmogenic Right Ventricular Cardiomyopathy   1
2. Congenital Heart Diseases   5
3. Constrictive Pericarditis   21
4. Hypertrophic Cardiomyopathy   35
5. Infective Endocarditis   38
6. Infiltrative Cardiomyopathy   50
7. Intracardiac Tumour *versus* Clot   57
8. Left Ventricular Aneurysm *versus* Pseudoaneurysm   80
9. Left Ventricular Assist Device   84
10. Myocardial Infarction (MI), Dysfunction and Coronary Artery Disease   87
11. Mediastinal and Other Mediastinal/Chest Wall Lesions   97
12. Vascular Anomalies and Diseases   115
    Index   185

List of Abbreviations

AE	Atrial enlargement		FDG-PET	Fluorodeoxyglucose-positron emission tomography
AF	Atrial fibrillation			
ARVC	Arrhythmogenic right ventricular cardiomyopathy		FSE	Fast spin echo
			Gd	Gadolinium
ARVD	Arrhythmogenic right ventricular dysplasia		GE	Gradient echo
AR	Aortic regurgitation		HCM	Hypertrophic Cardiomyopathy
AV	Atrioventricular			
			HE	Hyper eosinophilia
BPH	Benign prostatic hypertrophy		HES	Hyper eosinophilic syndrome
4CH	4 Chamber		HHE	Delayed high heterogeneous enhancement
CABG	Coronary artery bypass grafting			
			HRCT	High resolution computed tomography
CAD	Coronary artery disease			
CECT	Contrast enhanced computed tomography		IVC	Inferior vena cava
			IV	Intravenous
CHD	Congenital heart disease			
CFA	Common femoral artery		LA	Left atrium
CKD	Chronic kidney disease		LAD	Left anterior descending artery
CMR	Cardiovascular magnetic resonance			
			LGE	Late gadolinium enhancement
CT	Computed tomography			
CTA	Computed tomography angiography		LRTI	Lower respiratory tract infection
			LV	Left ventricle
DE-MRI	Delayed-enhancement MRI		LVA	Left ventricular aneurysm
DIR	Double inversion recovery		LVAD	Left ventricular assist device
DM/HTM	Diabetes mellitus and associated hypertension		LVEDV	Left ventricular end-diastolic volume
ECG Gating	Electrocardiographic gating		LVESV	Left ventricular end-systolic volume
EFE	Endocardial fibroelastosis			
EMF	End myocardial fibrosis		LVEF	Left ventricular ejection fraction
EPS	Electrophysiology study			
EVAR	Endovascular aneurysm repair/endovascular aortic repair		LVH	Left ventricular hypertrophy
			LVOT	Left ventricular outflow tract
			LVT	Left ventricular thrombi

MAPCAs	Multiple aorto-pulmonary collaterals	SAX	Short axis
MHE	Moderate heterogeneous enhancement	SDFP	Steady state free precession
		STIR	Short-tau inversion-recovery
MR	Mitral regurgitation	SVC	Superior vena cava
MRA	Magnetic resonance angiography	SVG to OM	Saphenous venous graft to obtuse marginal
MVR	Mitral valve replacement	TA	Takayasu arteritis
PC-MRI	Phase-contrast MRI	TAA	Thoracic aortic aneurysm
PE	Pericardial effusion		
PF	Papillary fibroelastoma	TIR	Triple inversion recovery
PR	Pulmonary regurgitation	TB	Tuberculosis
PTCA	Percutaneous transluminal coronary angioplasty	TEE	Transesophageal echocardiography
PTFE	Polytetrafluoroethylene	TOF	Tetralogy of Fallot
PV	Pulmonary valve	TR	Tricuspid regurgitation
RA	Right atrium	TSE	Turbo spin echo
RCA	Right coronary artery	TTE	Transthoracic echocardiography
RCM	Restrictive cardiomyopathy		
RV	Right ventricle	VF	Ventricular function
RVM	Right ventricular myocardium	VH	Ventricular hypertrophy
		VSD	Ventricular septal defect
RVOT	Right ventricular outflow tract	VT	Ventricular tachycardia

Protocols

GENERAL MR PROTOCOL

All the datasets were obtained on a 3-T scanner, General Electric Medical Systems and included both fast spin-echo (FSE) and gradient-echo (GE) sequences.

MR imaging examinations were performed by using a dedicated cardiac coil. Images were acquired with electrocardiographic gating (ECG gating) and a steady-state cine technique (fast imaging employing steady-state acquisition) and with 3.7–7.1/1.4–3.1 (repetition time msec/echo time msec), in short-axis oblique, long-axis oblique, and transverse planes. This was followed by perfusion studies and delayed contrast-enhanced imaging. Perfusion studies were performed by using previously described methods (8) during intravenous injection of 10 ml of gadolinium at a dose of 0.5 mmol/kg and a rate of 4 ml/sec with an automatic injector. Delayed contrast-enhanced images were obtained after an additional injection of 0.2 mmol/kg gadolinium at a volume of up to 20 ml by using an inversion-recovery-prepared breath-hold gradient-echo cine imaging sequence (5.4–7.1/1.4–3.1/200 [repetition time msec/echo time msec/inversion time msec]; flip angle of 25°; section thickness of 10–12 mm). Long-axis and short-axis oblique views were obtained 15 and 20 minutes, respectively, after the contrast agent injection. In four patients, additional contrast-enhanced fast steady-state cine images were acquired in selected planes as clinically indicated during the examination.

SPECIAL MR PROTOCOL

Quantitative evaluation of ventricular function (VF) is achieved by obtaining a series of contiguous cine steady state free precession (SSFP) slices that cover the ventricles in the short-axis plane. VEC MRI flow measurements are acquired to measure cardiac output, pulmonary-to-systemic flow ratio and for quantification of valve regurgitation.

An axial stack of contiguous cine SSFP slices covering the thorax from the hepatic inferior vena cava (IVC) through the aortic arch provides a comprehensive high-resolution dynamic survey of the thoracic anatomy. Additional targeted cine SSFP stacks (often with thinner sections for improved spatial resolution) in oblique planes can then be prescribed across areas of interest to enhance interpretation. Turbo (fast) spin echo with blood suppression (TSE) can be utilized in a similar fashion and is less affected by image artifact from implanted metal devices. Although TSE provides high resolution, high contrast images, it has the disadvantage of providing only static (non-cine) images. Another useful technique to visualize the thoracic vasculature is gadolinium (Gd)-enhanced 3D magnetic resonance angiography (MRA).

T2-weighted spin-echo CMR, preferably using a short-tau inversion-recovery (STIR) sequence (also called "triple-inversion" spin-echo), highly useful to depict myocardial edema, can also be applied to detect pericardial fluid and/or edema of the pericardial layers in patients with inflammatory pericarditis. Use of paramagnetic contrast agents can be recommended in case of pericardial masses, inflammatory pericarditis, to depict concomitant myocardial pathology (e.g. myocarditis) and may be useful to better differentiate between inflammatory and constrictive forms of pericarditis.

Cardiac MR tagging was performed with saturation recover preparatory pulse.

- It is an improved T_1 weighting
- Applying a 90° flip angle
- Less subject to signal intensity variations than IR techniques.
- The saturation recover pulse is often used in combination with GRE and hybrid GRE–EPI pulse sequence.

CT Angiography Protocol

CT scan was done on multidetector scanner, Philips (1). Plain CT scans of thorax followed by (2). Early arterial phases: Was obtained after negative test dose of the non-ionic contrast by injecting 12 to 40 ml of intravenous contrast (Omnipaque) through cubital vein by pressure injector at a rate of 1.5 ml/sec. Scan was obtained 6 sec after starting of the intravenous contrast. Scan parameters were as follows:

Volumetric data was obtained from the vessels in axial plane and was reconstructed in sagittal, coronal plane. Slice thickness: –1 mm. Collimation: –0.6 mm. Pitch: –1.5, MAS = 200, Kvp = 120.

CT Angiography Protocol for LVAD

Retrospectively gated contrast-enhanced CTA was performed on 64-detector scanner, and the CTA images were post-processed in multiple curved projections on workstation. A comprehensive protocol to evaluate an LVAD would include ECG gating, intravenous contrast material, and coverage from the aortic arch (to include the outflow cannula) through the abdomen (to include the pump and the exit of the power cord through the skin).

CT Angiography Protocol for Aortic Dissection

A typical helical scanning protocol for aortic dissection includes the following parameters: 5 mm collimation, 1.5 pitch, and 7.5 mm imaging spacing. Multidetector CT may be performed with 1–2.5 mm collimation. Initial nonenhanced CT is used for the diagnosis of acute haemorrhage and aortic rupture. This is followed by helical CT performed approximately 25–30 seconds after the injection of contrast material. Non-ionic contrast material (120–135 ml) is power injected via a peripheral intravenous (IV) site at a rate of 3–4 ml/sec.

Arrhythmogenic Right Ventricular Cardiomyopathy

Arrhythmogenic right ventricular cardiomyopathy (ARVC), also known as arrhythmogenic right ventricular dysplasia (ARVD), is characterized by progressive fibro fatty replacement of the right ventricular myocardium (RVM).

Case 1

Patient is 46-year-old male having vague chest discomfort. Holter was done which showed ventricular tachycardia (VT) arising from RVOT. He underwent echocardiography which showed dilated right ventricle (RV) and RV systolic dysfunction. Patient was advised for cardiac MRI.

Cardiac MRI Findings

Right Ventricle

- Dilated right atrium (RA) and RV measuring 4.3 cm in dimension (Figs 1.1a and b).
- RV systolic dysfunctions were—RV EF (ejection fraction) = 30% and a small aneurysmal bulging arising from anterior aspect of RVOT (Fig. 1.2).
- RV free wall is thin, shows fatty infiltration as noted in double inversion recovery (DIR) and tripple inversion recovery (TIR) (Figs 1.3a and b).
- Delayed enhancement sequence showed enhancement in the free wall (Fig. 1.4)

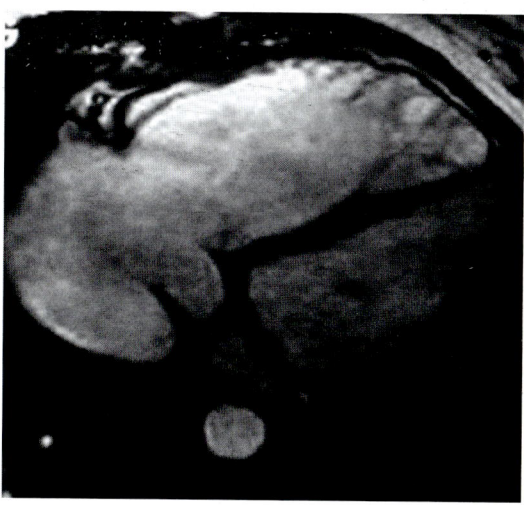

Fig. 1.1 a: Cardiac MRI four-chamber image showing dilated RA and RV

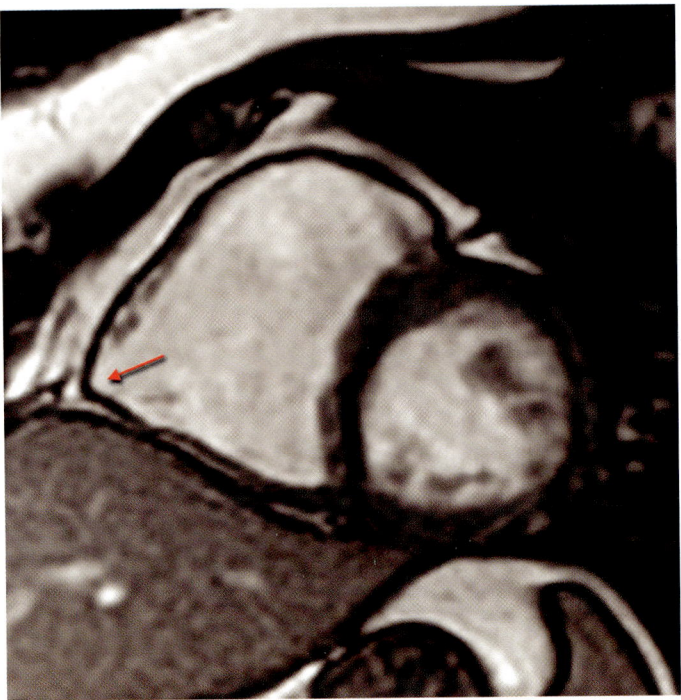

Fig. 1.1 b: Cardiac MRI image showing dilated right ventricle with relative thin free wall (arrow).

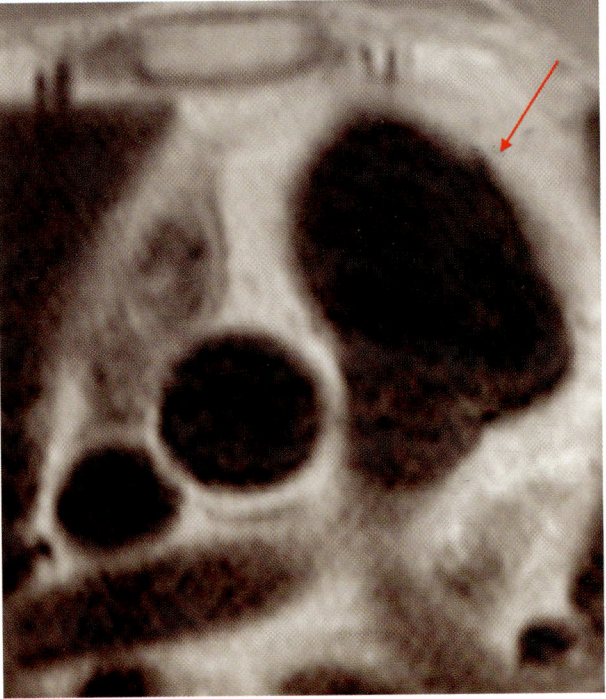

Fig. 1.2: Cardiac MRI showing small bulging in RVOT (arrow)

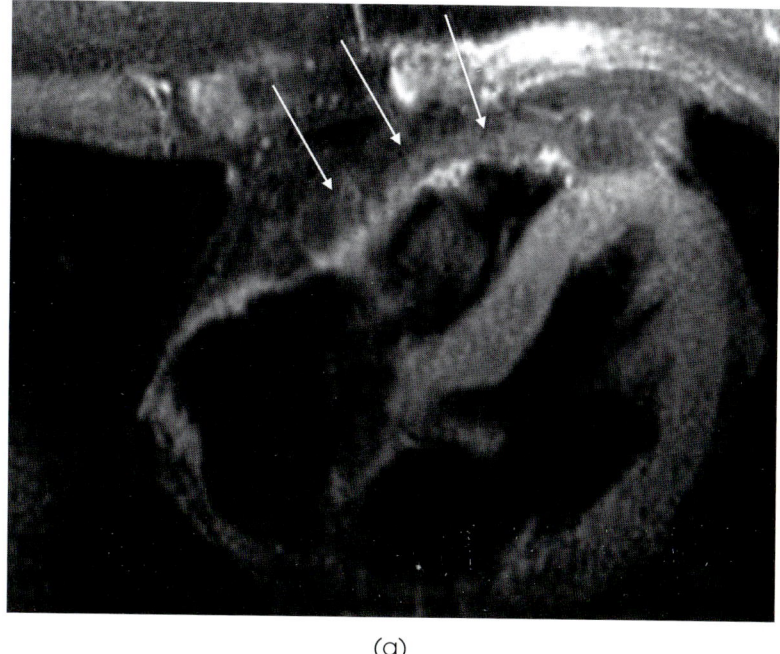

(a)

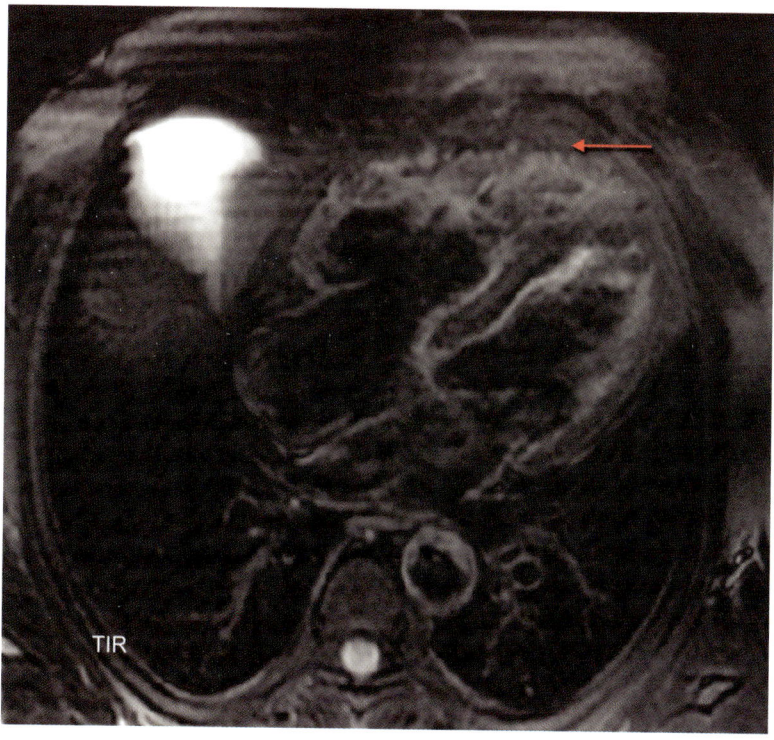

(b)

Fig. 1.3: DIR and TIR image showing fatty deposition in RV wall (arrows)

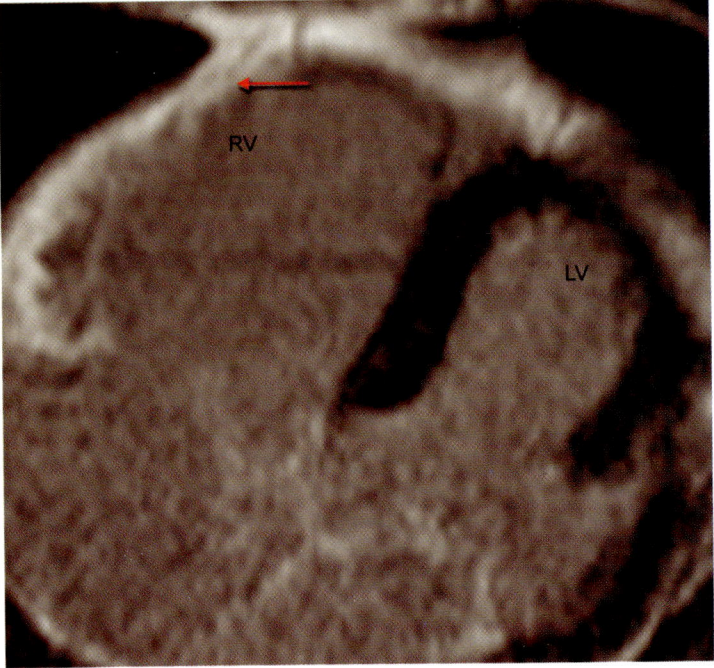

Fig. 1.4: Enhancing RV wall (arrow)

Left Ventricle

- Normal LA (left atrium) and LV (left ventricle) size
- Normal LV systolic function.

Diagnosis

ARVD (Arrhythmogenic right ventricular dysplasia).

Surgical /Medical Data

Medical Management

After electrophysiology study (EPS) the patient was kept on defibrillator.

Discussion

ARVD is not common but can be seen more frequently in young adults, and clinical manifestations range from no symptoms to lethal arrhythmia and sudden death. The diagnosis of ARVC is challenging and is based on the recently revised international task force criteria. Given the strengths of cardiac magnetic resonance (MR) imaging for depicting the RV, this modality plays an important role in the diagnosis of ARVC. Functional and structural abnormalities of the RV depicted with cardiac MR imaging constitute major and minor criteria in the revised task force criteria.

Suggested Reading

Corrado D, Basso C, Thiene G. Arrhythmogenic right ventricular cardiomyopathy: an update. Heart 2009; 95: 766–773.

Congenital Heart Diseases

Patient is 36-year-old female, status postoperative tetralogy of Fallot (TOF) in the past, now complaints of exertional dyspnoea since one month. Cardiac MRI was done for further evaluation.

Cardiac MRI Findings

- Dilated right ventricle (RV) and RVOT (Figs 2.1 and 2.2)
- Normal right ventricular (RV) systolic function and normal calibre pulmonary arteries (Figs 2.3a and b)
- Severe pulmonary regurgitation (PR) (Fig. 2.4)
- Mild tricuspid regurgitation
- Normal left ventricular (LV) systolic function
- No enhancement in RV wall (Fig. 2.5).

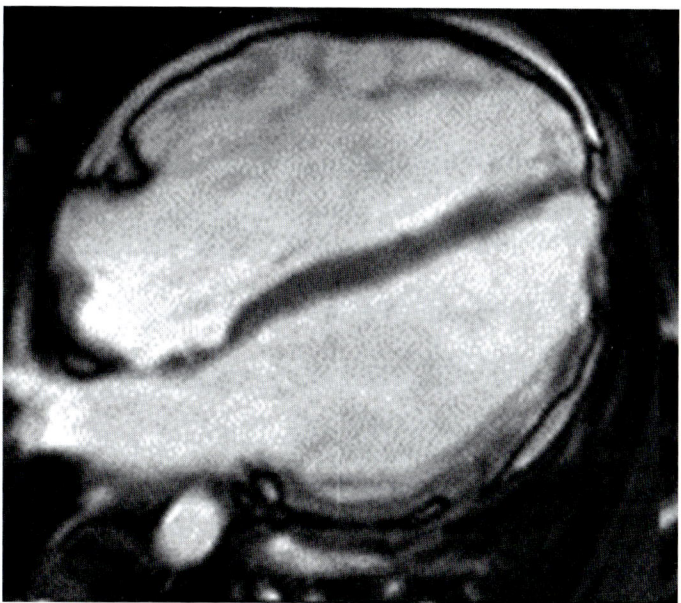

Fig. 2.1: MRI showing dilated right ventricle

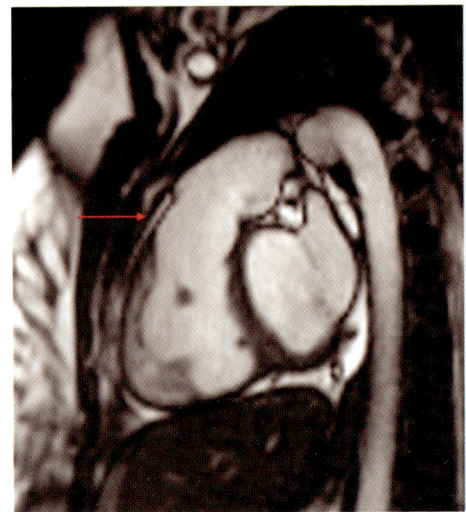

Fig. 2.2: MRI RVOT image showing dilated RVOT (arrow)

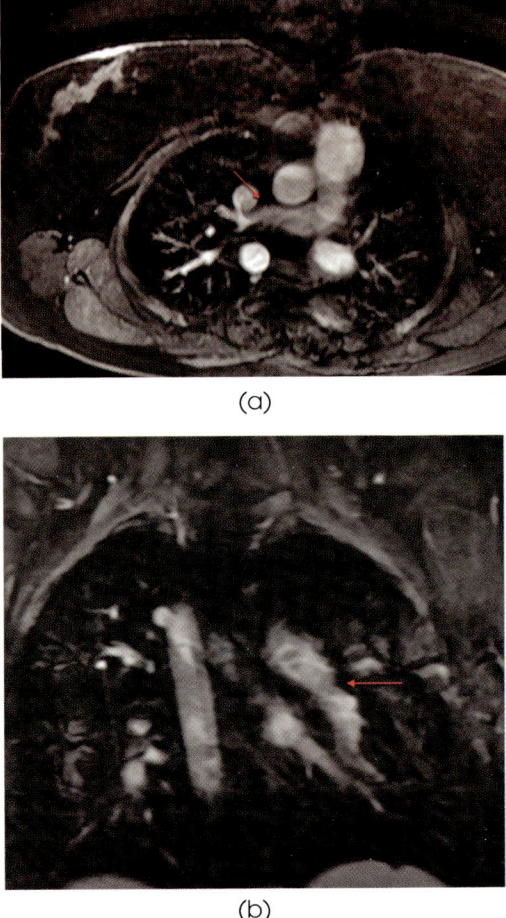

(a)

(b)

Fig. 2.3a and b: MRI showing normal calibre pulmonary arteries (arrows).

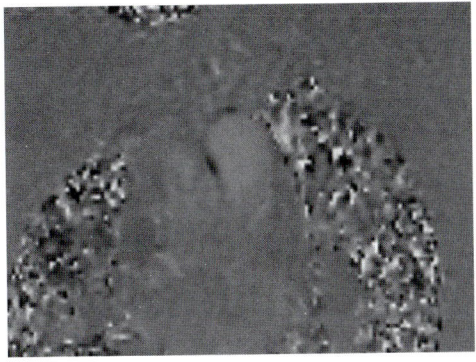
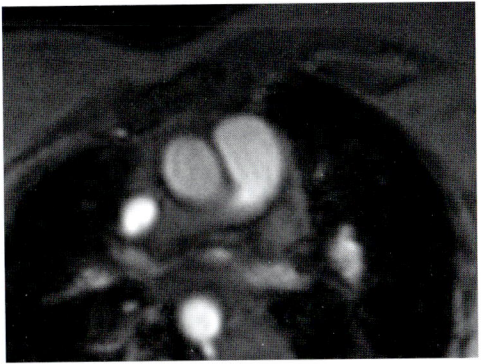

Fig. 2.4: Phase-contrast MRI (PC-MRI) for pulmonary flow analysis

- Peak positive velocity—7 cm/sec
- Peak negative velocity—51.5 cm/sec
  Pulmonary flow analysis showed severe pulmonary regurgitation

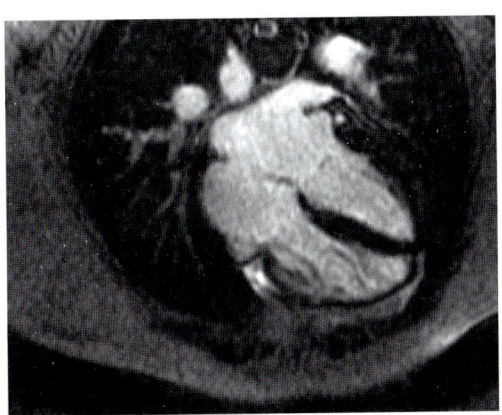
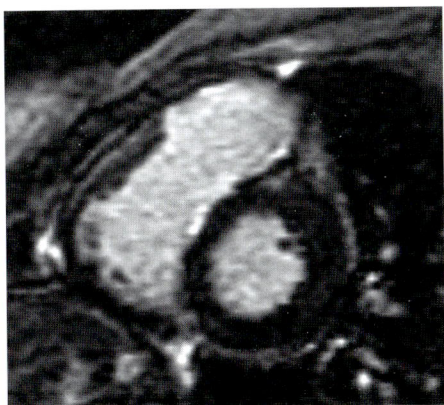

Fig. 2.5: MRI showing no delayed enhancement in right or left ventricular wall

RV Function

- RVEDV—72 ml, ESV—35 ml
- RV EF—51%

Radiological Diagnosis

Pulmonary regurgitation in postoperative TOF REPAIR.

Treatment

Percutaneous pulmonary stent placement was done.

Discussion

Adults with a history of TOF should be regularly monitored with noninvasive imaging for evidence of deterioration of cardiac function, even before the onset of symptoms. Proper timing of surgical interventions, such as PV replacement, will allow maximum recovery of cardiac function. Because of its wide availability and high utility, 3D VR multidetector CT is being used with increasing frequency to assess this patient population, especially in cases in which MR imaging is contraindicated or susceptible to imaging artifacts. Diagnostic evaluation requires familiarity with the spectrum of MR appearances described in this article, as well as knowledge of the anatomy, associated defects, natural history, surgical management, and common postoperative complications of TOF. With ongoing technologic advancements, the role of MR in clinical practice will continue to expand, and this modality will contribute significantly to decisions regarding the care of patients who have undergone TOF repair.

Suggested Reading

Norton KI, Tong C, Glass RB, Nielsen JC. Cardiac MR imaging assessment following tetralogy of Fallot repair. Radiographics 2006; 26(1): 197–211.

Case 2

Patient came with chief complaints of cyanosis since birth progressively increasing. Repeated lower respiratory infection (LRTI) since birth. X-ray chest was done showed dilated RA. Echocardiography was done, then patient was advised for cardiac MRI for further evaluation.

Cardiac MRI Findings

- Situs solitus, levocardia.
- Morphological RV is dilated. Large atrioventricular (AV) defect (Fig. 2.6).
- Atrioventricular and ventriculo-atrial discordance (Figs 2.7a and b).
- Normal pulmonary venous anatomy (Fig. 2.8).
- Morphological left ventricle is lying right to morphological right ventricle, so morphological right atrium is connected to morphological left ventricle and morphological left atrium to morphological RV (Fig. 2.9).
- Pulmonary artery to morphological LV and lies posterior to right of aorta. Aorta is left sided (Fig. 2.9).
- Severe pulmonary valvular atresia and thickened pulmonary valve. Valvular region—11.7. Supravalvular—13 mm (Fig. 2.10).
- Confluent pulmonary branches.
- Normal aortic branches (Fig. 2.11).
- Noninterrupted IVC (Fig. 2.12).
- Single superior vena cava (SVC) (Fig. 2.12).

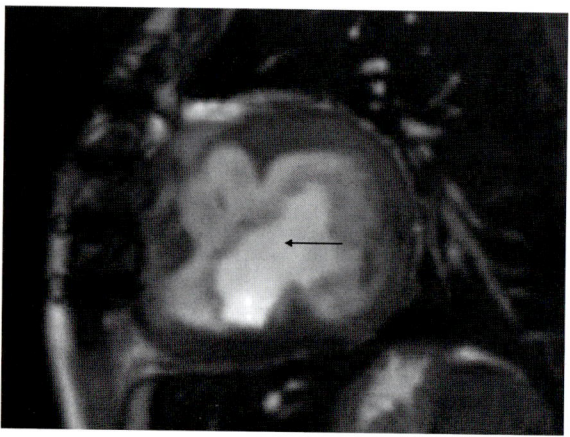

Fig. 2.6: Cardiac MRI showing large AV canal defect (arrow)

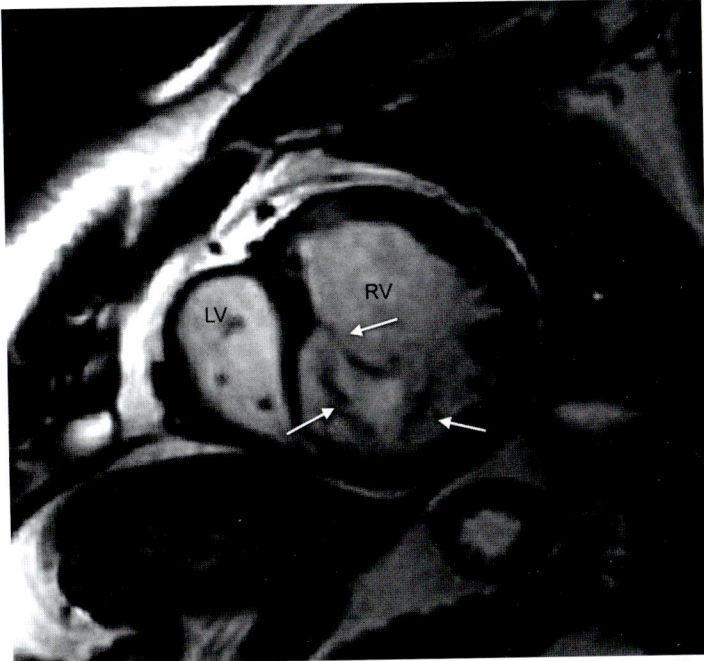

Fig. 2.7a: MRI showing opening of mitral valve in RV (arrows)

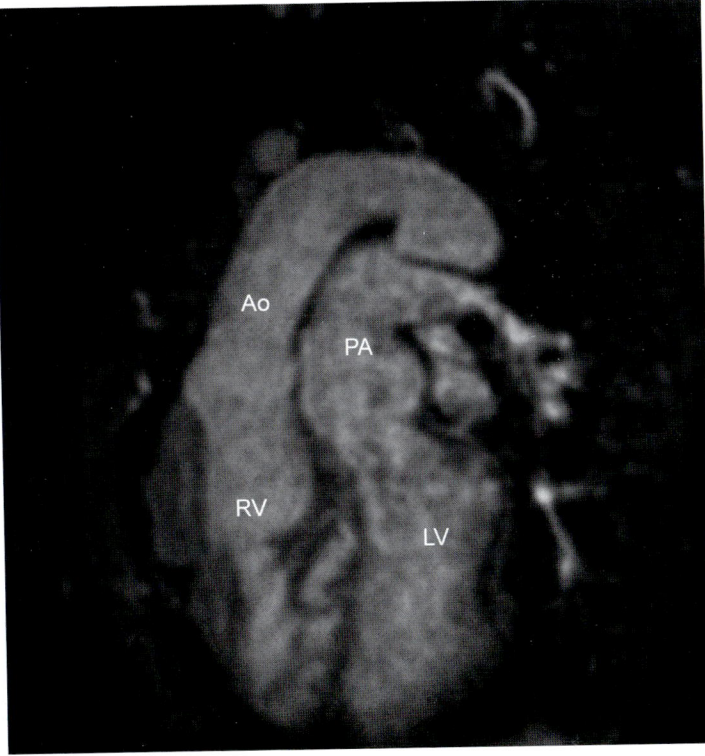

Fig. 2.7b: MRI showing aorta taking origin from RV and pulmonary artery from LV. Aorta is anterior to pulmonary artery

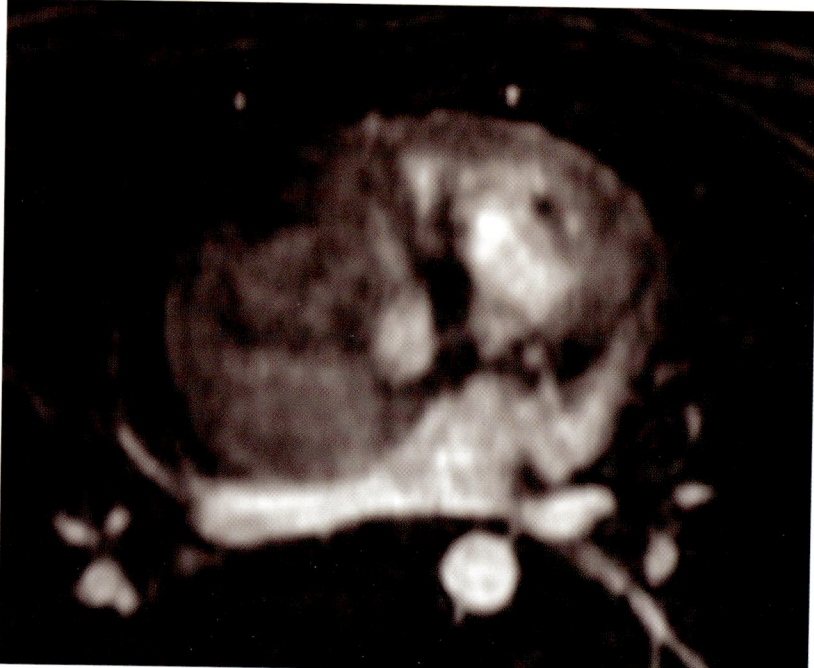

Fig. 2.8: MRI showing RT and left pulmonary veins

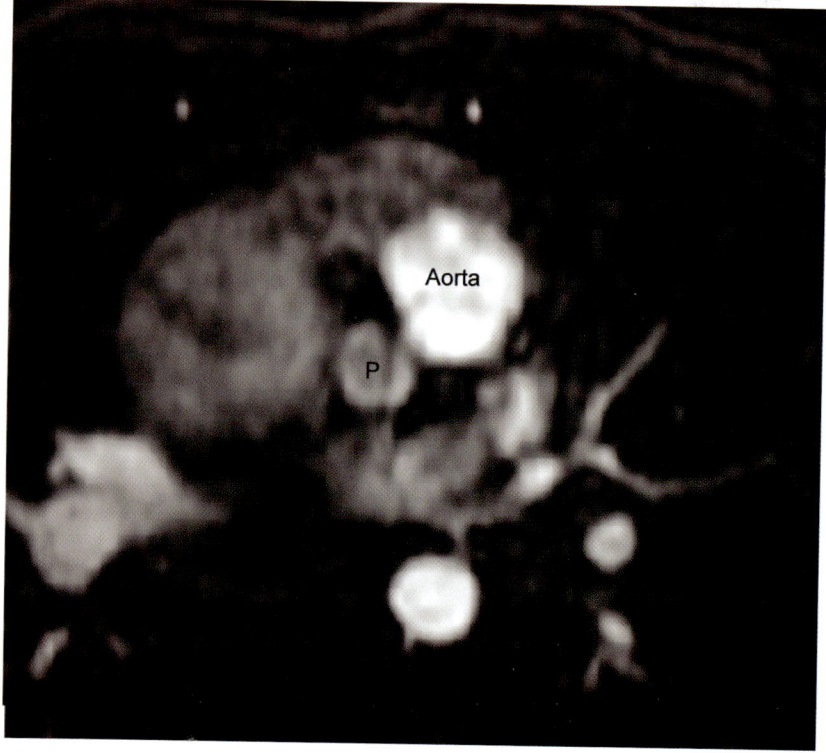

Fig. 2.9: MRI showing aorta is anterior and towards left side

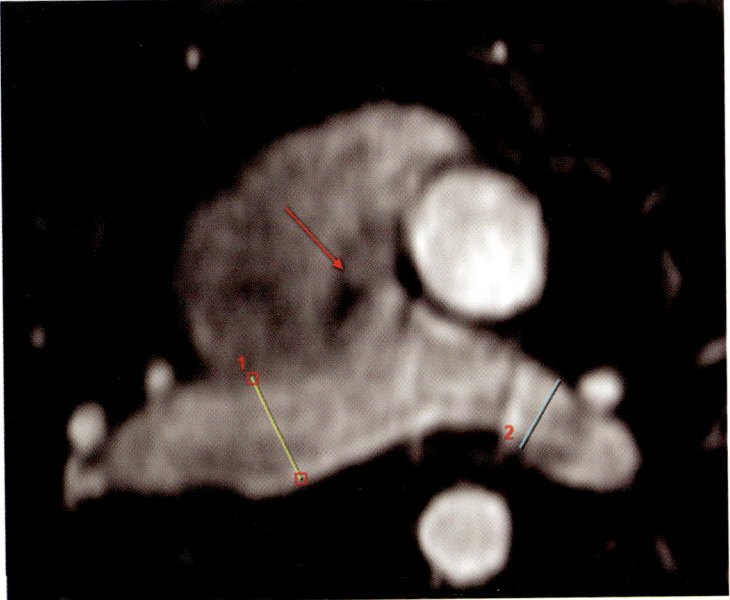

Fig. 2.10: MRI showing Infundibular atresia (arrow), RT and left pulmonary arteries

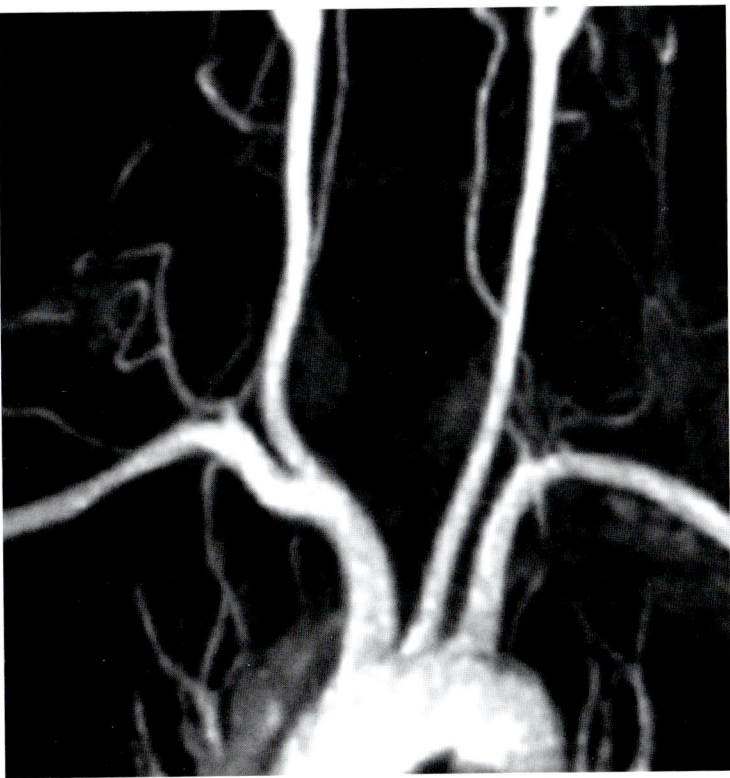

Fig. 2.11: MRI showing normal aortic branches

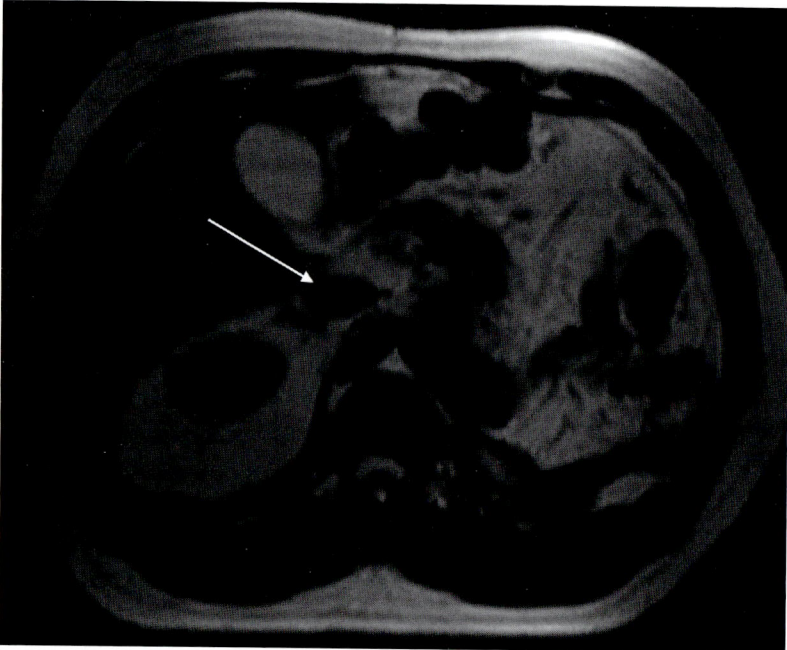

Fig. 2.12: MRI showing IVC (arrow)

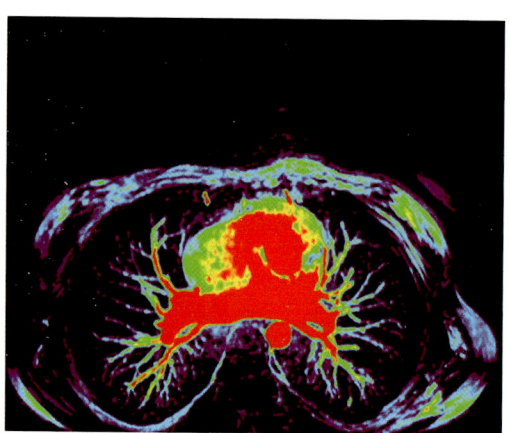

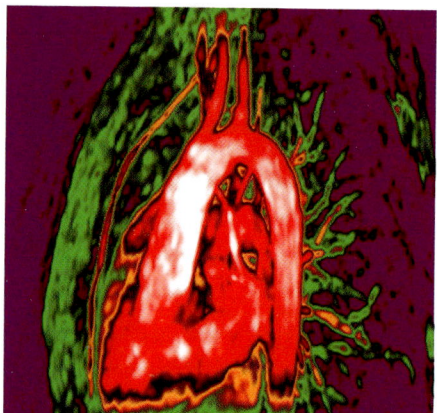

Fig. 2.13: Reformatted images

LV Function—EDV—34.7 ml, ESV—7.5 ml, EF—76%, SV—27.2 ml

Reformatted images showing same (Fig. 2.13).

Radiological Diagnosis

1. Corrected transposition of great vessel with atrio-ventricular and ventricular discordance.
2. Severe pulmonary valvular stenosis with dysplastic thickened pulmonary valve.
3. Pulmonary artery lying posterior and to the right side of aorta
4. Border line normal left and right ventricular function.

Diagnosis

CHD and c-TGA, VSD with severe pulmonary atresia, straddling left AV valve.

Surgical intervention

Extra cardiac fontan operation: Nonfenestrated extra cardiac fontan (20 mm PTFE tube) + MPA ligation + atrial septectomy

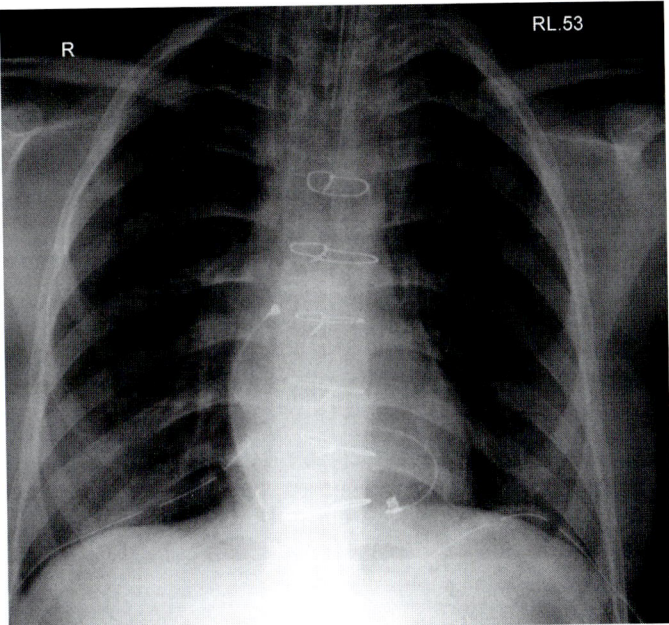

Fig. 2.14: Postoperative CXR 5th day

Discussion

CMR is a very useful tool in the diagnosis, evaluation and management of patients with congenital heart disease (CHD). For CHD patients, the whole spectrum of CMR techniques is employed and delivers valuable information concerning ventricular and valve function, anatomy of malformed, surgically or interventionally corrected or palliated hearts and great vessel defects. It helps to correlate exercise capacity and prognostic factors such as scar formation, fibrosis or ventricle size. Depending on the individual defect, CMR complements other imaging modalities such as echocardiography and heart catheterization.

New MR techniques are being developed that in future might help to increase the role of CMR in the clinician's decision making. Among these are real time MRI, 3D-/4D-flow measurements, virtual surgery based on CMR data and the MRI heart catheterization laboratory. MRI based catheter interventions and measurements are possible in experimental settings. A future aim is the use of the advantages of CMR (high resolution, 3 dimensionality, and non-X-ray) and combine them with those of the catheterization laboratory (measurement of pressures, oxygen saturation, interventions).

Overall, CMR already does play an ever-increasing role in the management of CHD patients.

Suggested Reading

1. Gaca AM, Jaggers JJ, Dudley LT, Bisset GS 3rd (2008) repair of congenital heart disease: a primer part 1. Radiology 247: 617–631.
2. Gaca AM, Jaggers JJ, Dudley LT, Bisset GS 3rd (2008) repair of congenital heart disease: a primer part 2. Radiology 248: 44–60.

Case 3

A 3-year-old female child first in birth order, born out of a non-consanguineous marriage via normal vaginal delivery in a hospital.

She presented with a history of easy fatigueability and dyspnoea on exertion since two and a half months of life. She also presented with a history of recurrent respiratory tract infections since birth. She was taken to a local paediatrician for a medical opinion when he noted clubbing and cyanosis. Her echo was done which revealed ventricular septal defect (VSD) with pulmonary atresia and multiple aorto-pulmonary collaterals (MAPCAs). The child was then referred to a higher centre for further management. CT angiography was done.

CT Angiography

- VSD (Fig. 2.15)
- Situs solitus, levocardia
- Left-sided aortic arch with normal branches
- Infundibular valvular and supravalvular atresia with narrowed confluent pulmonary branches (Fig. 2.16).
- RA and RV dilated
- PDA noted
- Few collaterals
- Normal pulmonary veins (Fig. 2.17)
- Aberrant right subclavian artery (Figs 2.18 to 2.20)

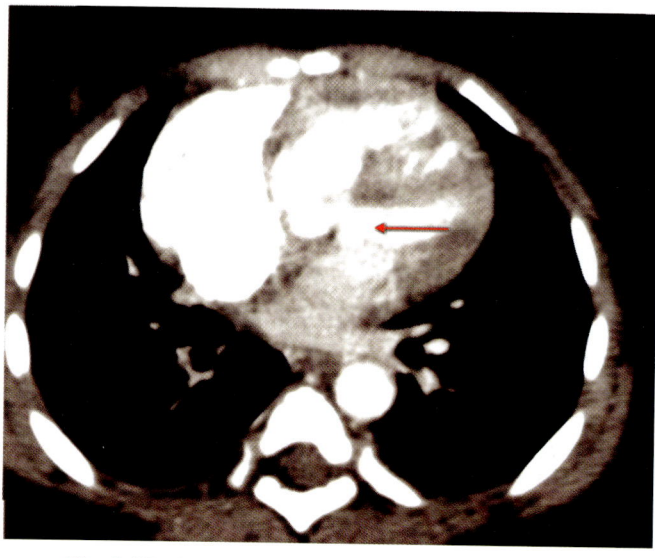

Fig. 2.15: CT angiography showing VSD (arrow)

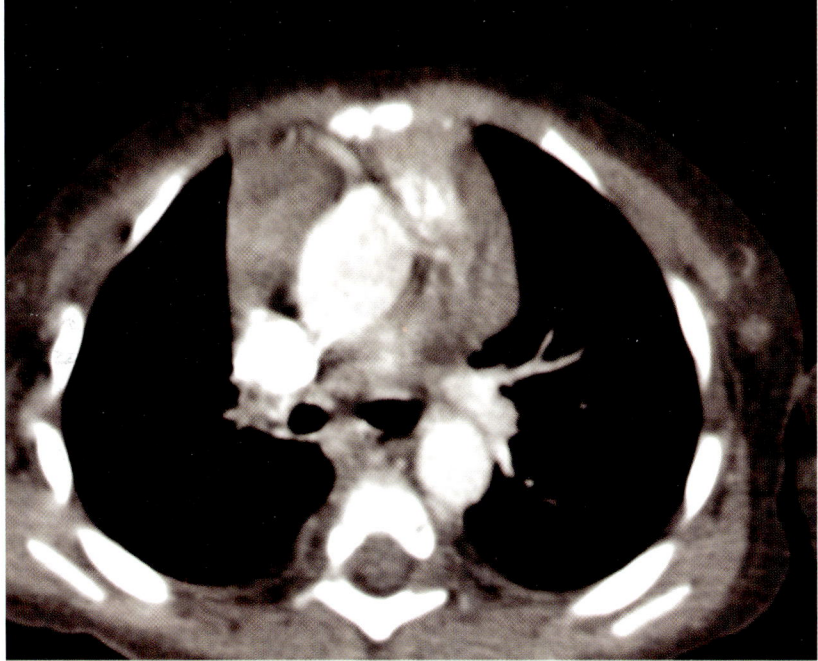

Fig. 2.16: CT showing atretic RVOT and narrow pulmonary branches

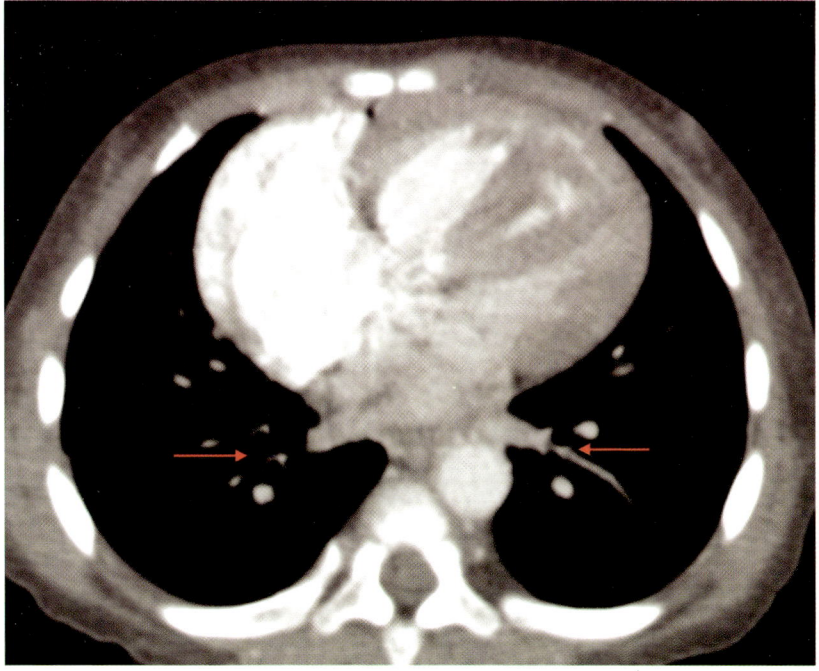

Fig. 2.17: CT showing normal pulmonary veins (arrows)

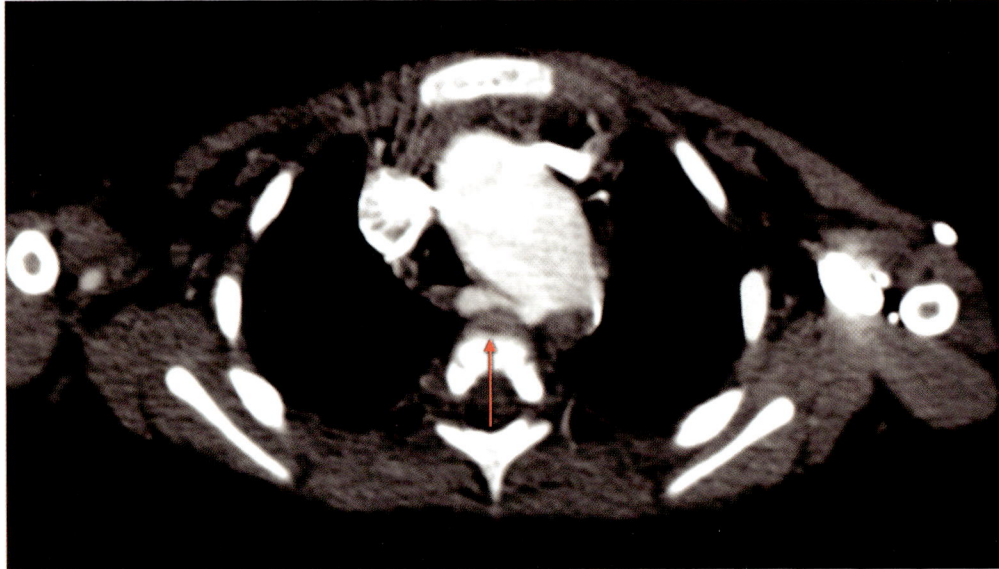

Fig. 2.18: CT showing aberrant RT subclavian artery (arrow)

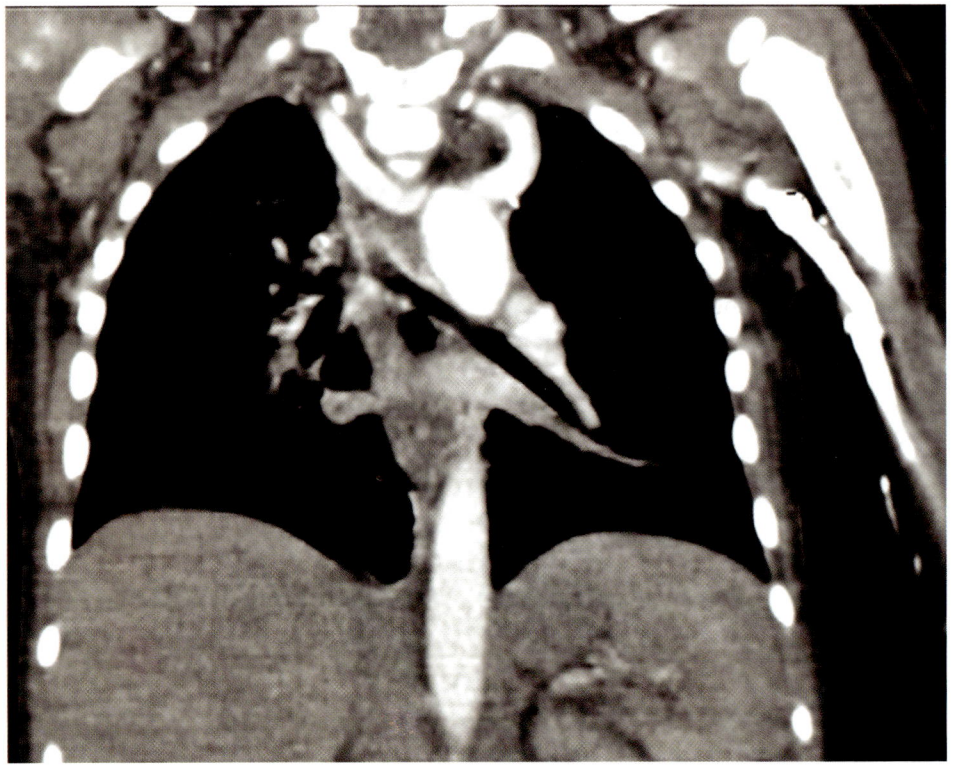

Fig. 2.19: CT coronal reformatted image showing aberrant RT subclavian artery

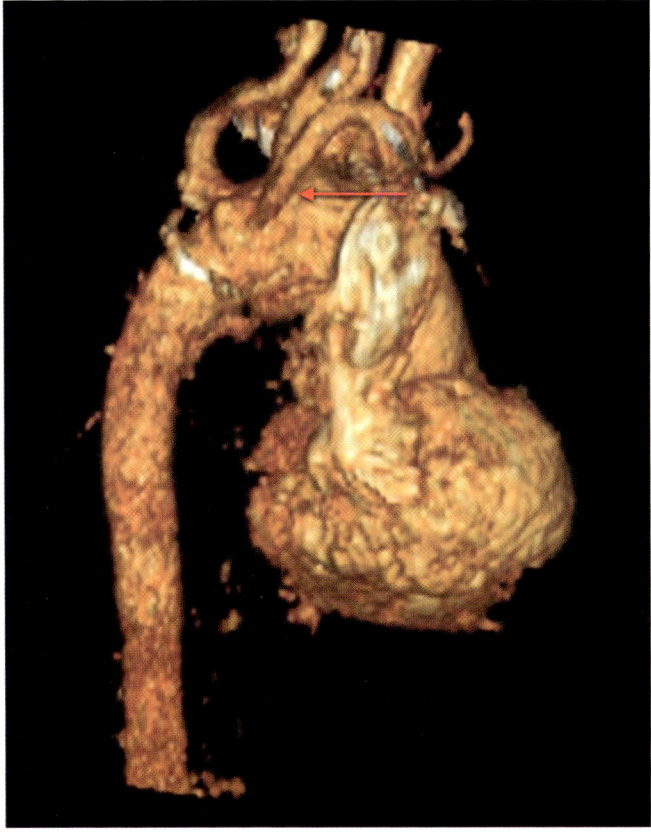

Fig. 2.20: CT volume rendering image showing aberrant origin of RT SCA (arrow)

Surgical Intervention

Surgery was done and patient was discharged successfully.

Discussion

Adults with a history of TOF should be regularly monitored with non-invasive imaging for evidence of deterioration of cardiac function, even before the onset of symptoms. Proper timing of surgical interventions, such as pulmonary valve (PV) replacement, will allow maximum recovery of cardiac function. Because of its wide availability and high utility, 3D VR multi-detector CT is being used with increasing frequency to assess this patient population, especially in cases in which MR imaging is contraindicated or susceptible to imaging artefacts. Diagnostic evaluation requires familiarity with the spectrum of CT appearances described in this article, as well as knowledge of the anatomy, associated defects, natural history, surgical management, and common postoperative complications of TOF. With ongoing technologic advancements, the role of CT in clinical practice will continue to expand, and this modality will contribute significantly to decisions regarding the care of patients who have undergone TOF repair.

Suggested Reading

1. Stinn B, Stoltzmann P, Fornaro J, et al. Technical principles of computed tomography in children with CHD. Insights Imaging. 2011; 2: 349–356.
2. Goo HW, Yang DH. Coronary artery visibility in free breathing young children with CHD on cardiac 64-slice CT: Dual-source ECG-triggered sequential scan vs. Single-source non-ECG-synchronized spiral scan. Paediatric Radiol. 2010; 40: 1670–1680.

Case 4

A 31-year-old female came with complaints of dyspnoea for last 3 years progressively increasing also gives history of syncopal attack. There was no history of diabetes/systemic hypertension /bronchial asthma/epilepsy/jaundice/lower limb claudication. No prior surgeries, no documented drug allergy/history of (h/o) myopia and family h/o Marfan syndrome.

On Examination

- All peripheral pulses well felt. No carotid/renal bruit
- Subluxation of lens
- Tall stature.

On echocardiography there was no regional wall motion abnormality, LVEF-60%, severe aortic regurgitation (AR), trace mitral regurgitation (MR), trace tricuspid regurgitation (TR). CT angiography was done.

CT Angiography Findings

Dilated aortic root and ascending aorta with normal aortic branches (Figs 2.21 to 2.23).

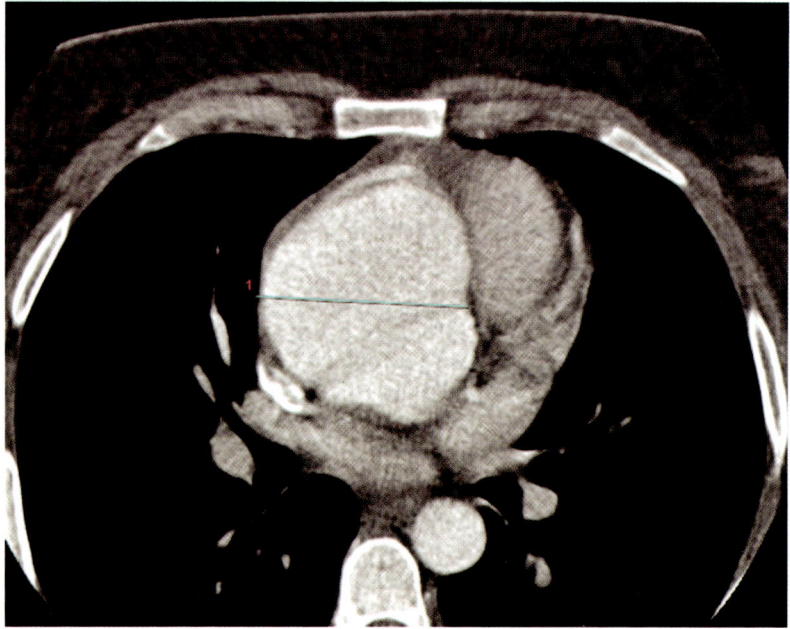

Fig. 2.21: CT angiography showing dilated ascending aorta

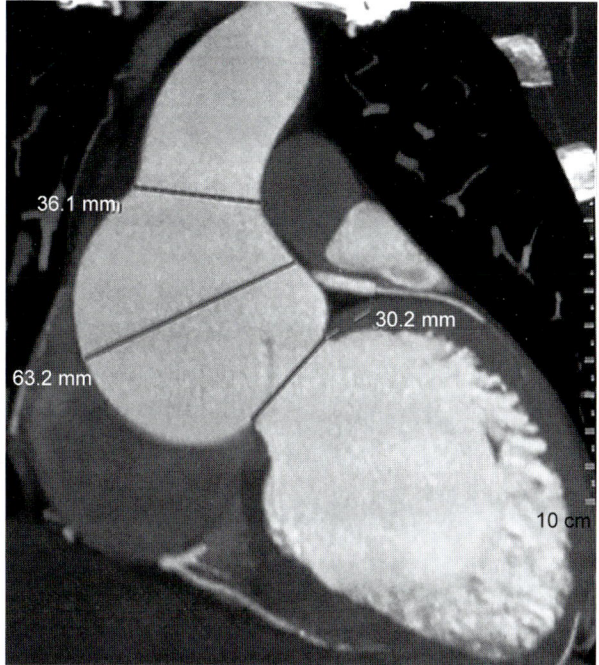

Fig. 2.22: CT angiography showing dilated root, sinus of Valsalva and ascending aorta

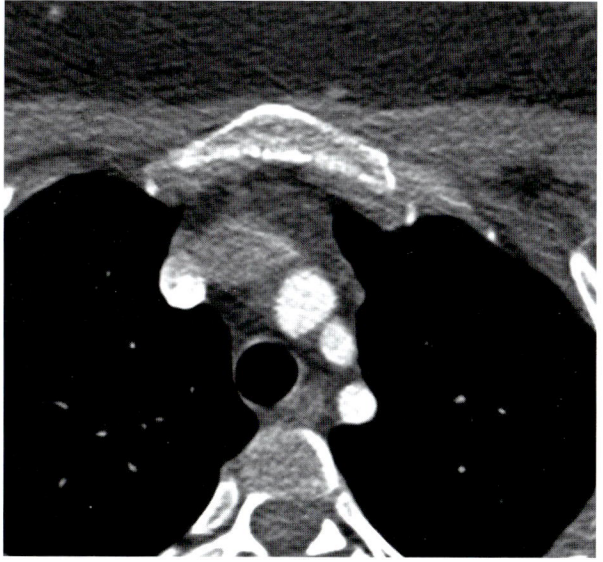

Fig. 2.23: CT showing normal aortic branches

Radiological Diagnosis

Marfan syndrome with dilated aortic root and ascending aorta, dislocation of lens, tall stature, h/o myopia and family h/o Marfan syndrome.

Surgical Intervention

Bentall's surgery was done as shown in Fig. 2.24.

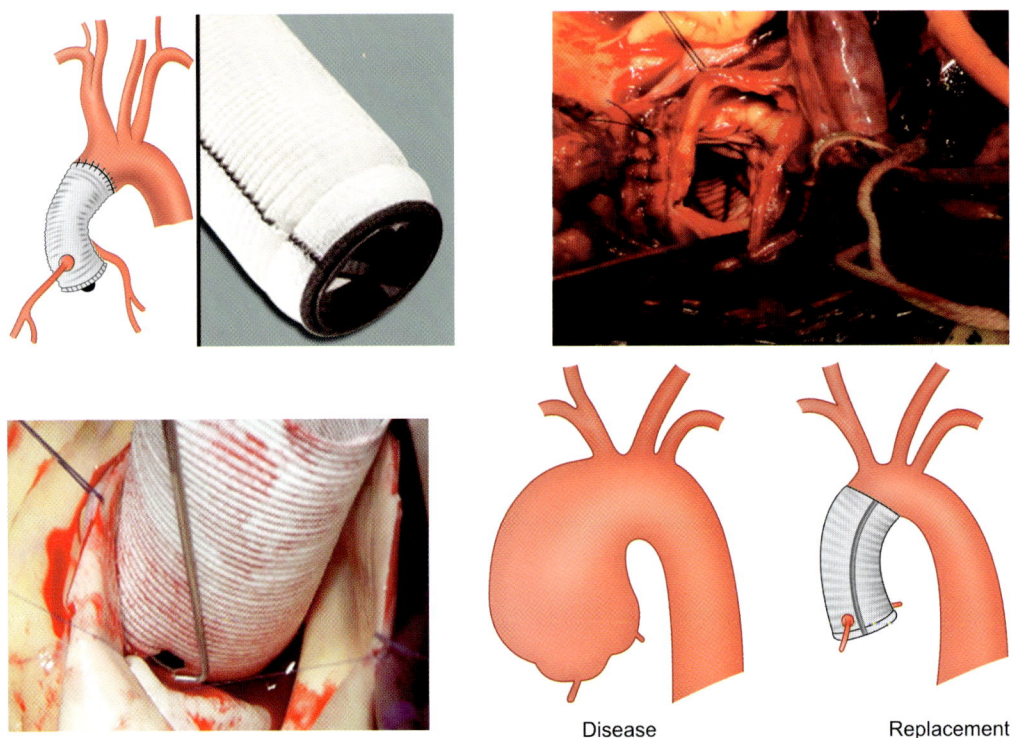

Disease   Replacement

Fig. 2.24: Bentall's surgery

Bentall's Surgery

Bentall's procedure is a cardiac surgical procedure involving composite graft replacement of the aorta valve, aortic root and ascending aorta with reimplantation of the coronary arteries into the graft.

Constrictive Pericarditis

Case 1

A patient came with history of breathlessness, pedal edema and ascites for two months. Patient has decreased urine output, with h/o weight loss for 1 month. Patient has h/o smoking and alcohol intake for past ten years.

Echocardiography was done which showed markedly irregular thickened pericardium with mild pericardial effusion (Fig. 3.1). There was septal bouncing with respiration. Doppler evaluation was done which was suggestive of constrictive pericarditis (Figs 3.2 and 3.3).

Patient was advised for cardiac MRI for further evaluation. Initially non-contrast CT scan was done to rule out calcification which showed extensive pericardial thickening. However, there was no evidence of pericardial calcification. B/L pleural effusion was noted (Fig. 3.4). Lung window showed left lung basal consolidation.

Cardiac MRI was done which showed following findings
1. Irregularly thickened pericardium measuring maximum up to approx. 25 mm at atrio-ventricular (AV) groove. There was minimal pericardial effusion.

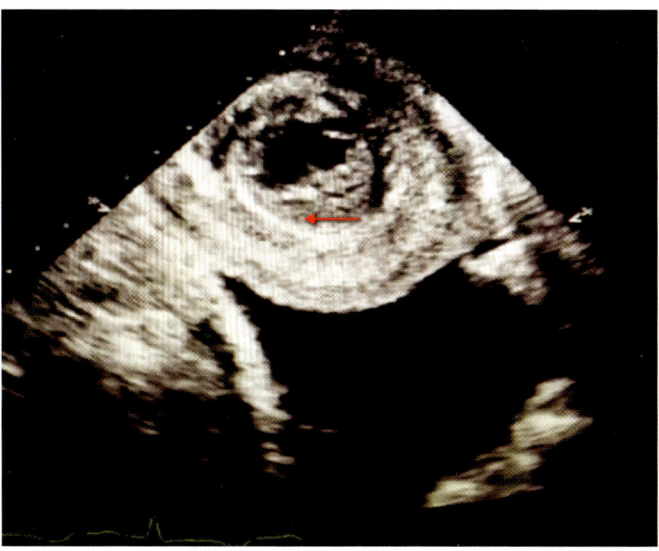

Fig. 3.1: Echocardiography showing thickened, irregular pericardium with mild pericardial effusion (arrow)

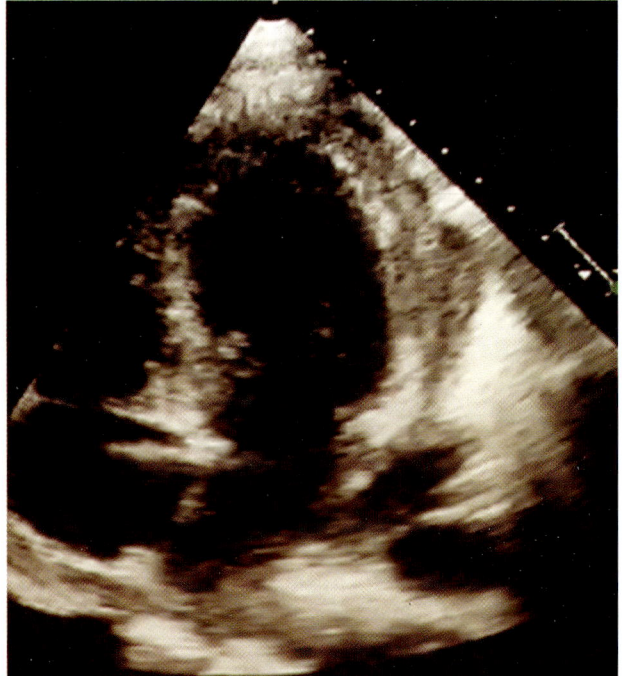

Fig. 3.2: Echocardiography

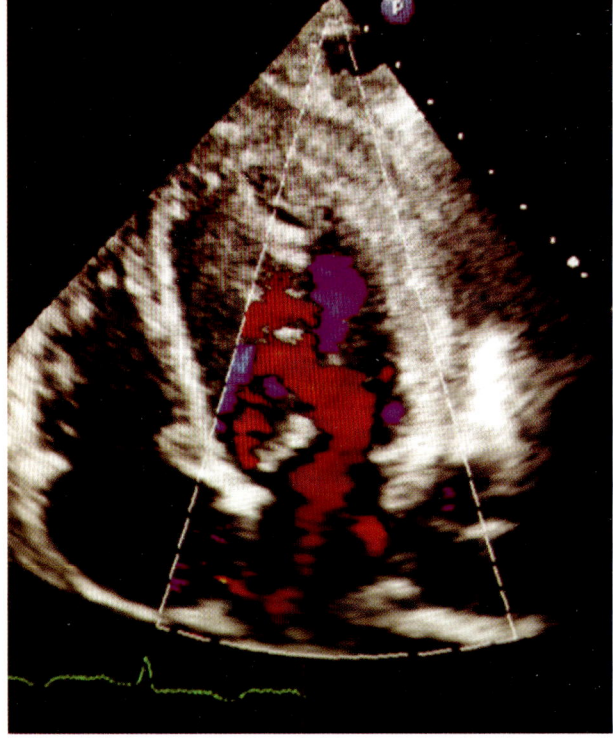

Fig. 3.3: Echocardiography

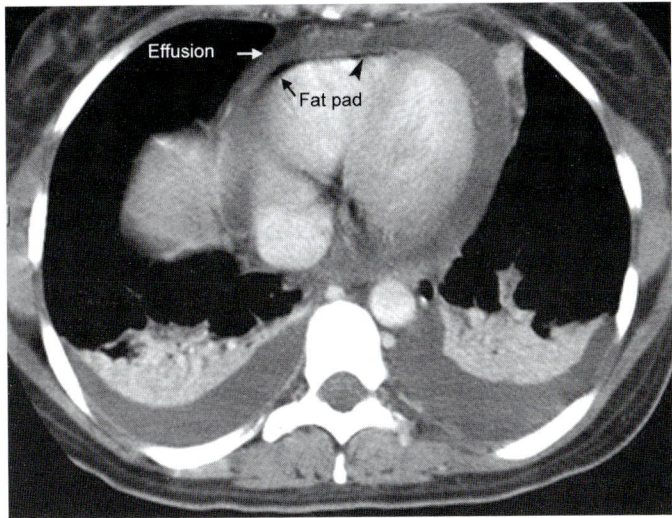

Fig. 3.4: CT scan showing pericardial thickening and B/L pleural effusion, pericardial effusion (arrow and arrowheads)

2. Contrast images showed diffuse homogenous enhancement of the pericardium. The thickened pericardium was adherent to myocardium. There was associated marked bilateral pleural effusion.
3. LV and RV regional function reveals reduced LV systolic function. Both RV and LV were relatively smaller in size and both atria were prominent. Cine images showed septal bouncing of LV septum. There was global hypokinesia of left and right ventricular wall.
4. Qualitative analysis showed reduced left ventricular systolic function. Overall features were suggestive of constrictive pericarditis. (Figs 3.5, 3.6 and 3.7). Patient then undergone pericardiectomy in our hospital. Pericardium was tubercular as suggested by histopathology.

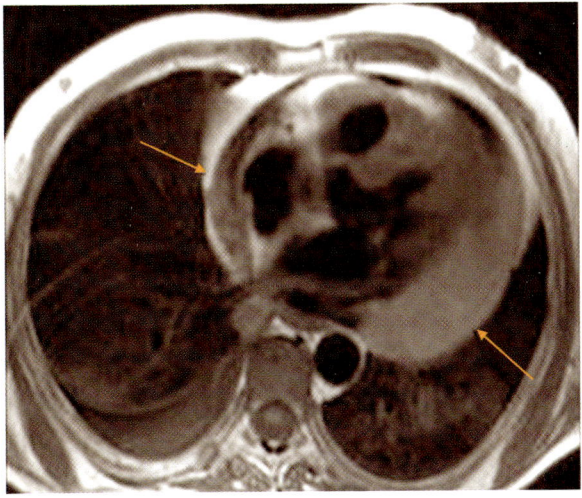

Fig. 3.5: Cardiac MRI (arrows)

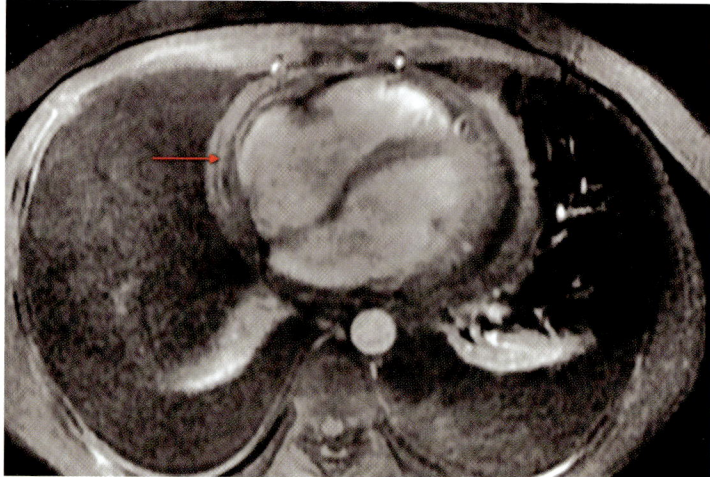

Fig. 3.6: Cardiac MRI (arrow)

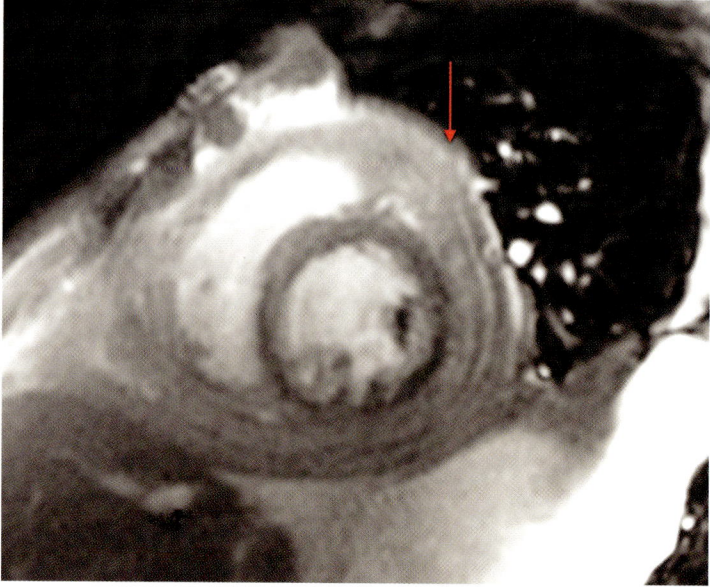

Fig. 3.7: Cardiac MRI (arrow)

Discussion

The pericardium and pericardial diseases in particular have received, in contrast to other topics in the field of cardiology, relatively limited interest. Today, despite improved knowledge of pathophysiology of pericardial diseases and the availability of a wide spectrum of diagnostic tools, the diagnostic challenge remains. Not only the clinical presentation may be atypical, mimicking other cardiac, pulmonary or pleural diseases; in developed countries a shift for instance in the epidemiology of constrictive pericarditis has been noted. Accurate decision making is crucial taking into account the significant morbidity and mortality caused by complicated pericardial diseases, and the potential benefit of therapeutic interventions. Echocardiography and cardiovascular magnetic resonance (CMR) are definitely the most versatile modalities to study the pericardium. It fuses excellent anatomic detail and tissue characterization with accurate evaluation of cardiac function and assessment of the haemodynamic consequences of pericardial constraint on cardiac filling. This review focuses on the current state of knowledge how CMR and echocardiography can be used to study the most common pericardial diseases.

Conclusion

Constrictive pericarditis is a potentially reversible cause of heart failure that may be difficult to differentiate from restrictive myocardial disease and severe tricuspid regurgitation (TR). MRI and echocardiography provide an important opportunity to evaluate for constrictive pericarditis, and definite diagnostic criteria are needed.

Echocardiography is very sensitive, specific, rapid and cost-effective non-invasive investigation for diagnosing pericardial effusion. Echocardiography may allow differentiation of constrictive pericarditis from heart failure due to restrictive myocardial disease or severe tricuspid regurgitation. Respiration-related ventricular septal shift, preserved or increased medial mitral annular E velocity, and prominent hepatic vein expiratory diastolic flow reversals are independently associated with the diagnosis of constrictive pericarditis.

The added value of CMR compared to the standard techniques used for assessment of patients with pericardial diseases has substantially increased in recent years. Strong points in favour of CMR are the integration of anatomic and functional information within a single examination, the ability for tissue characterization and to determine the presence and degree of inflammation and activity of disease, and the value of CMR to accurately assess the rest of the heart, in particular the myocardium, helpful in the differential diagnosis, which currently often remains a diagnostic challenge.

Suggested Reading

1. Kovanlikaya A, Burke LP, Nelson MD, Wood J. Characterizing chronic pericarditis using steady-state free-precession cine MR imaging. *Am J Roentgenol.* 2002; 179: 475–476.
2. Kojima S, Yamada N, Goto Y. Diagnosis of constrictive pericarditis by tagged cine magnetic resonance imaging. *N Engl J Med.* 1999; 341: 373–374.
3. Differentiation of constrictive pericarditis and restrictive cardiomyopathy by Doppler echocardiography. Circulation. 79 1989: 357–370.
4. Appleton CP, Hatle LK, Popp RL; Central venous flow velocity patterns can differentiate constrictive pericarditis from restrictive cardiomyopathy [abstract]. *J Am Coll Cardiol.* 9 1987:119A.

Case 2

Patient was known case of pulmonary tuberculosis (TB) and taking antitubercular treatment (ATT) and now advised for cardiac MRI to rule out pericardial involvement.

MRI Findings

1. Pericardium showing thickening measuring 4.8 mm at right atrioventricular groove and 4.6 mm at left atrioventricular groove (Fig 3.8).
2. Few mediastinal lymph nodes are seen in prevascular region

3. A well-defined focal lesion isointense to pericardium at level of free wall of right ventricle showing enhancement on post contrast study. The lesion was measuring 2.6 × 1.5 cm size. In view of patient history of pulmonary tuberculosis this lesion was considered as granulomatous lesion (Fig. 3.9).

4. CT scan showed small lesion isodense to pericardium at free wall of right ventricle (Fig. 3.10).

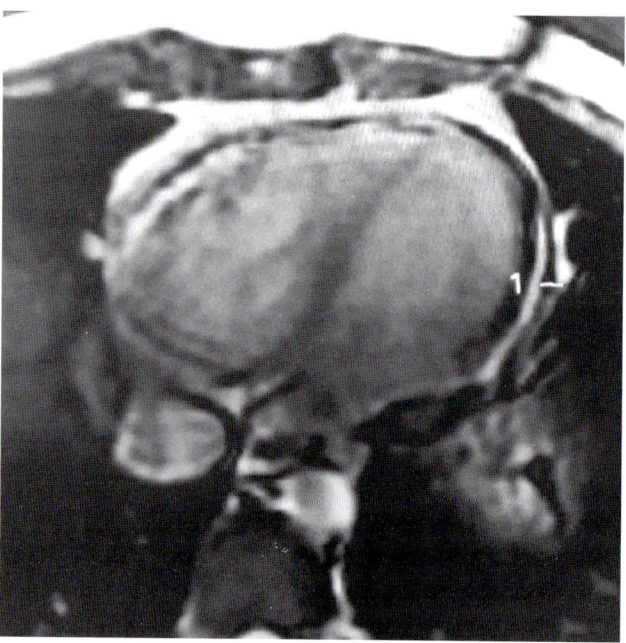

Fig. 3.8: Showing thickened pericardium

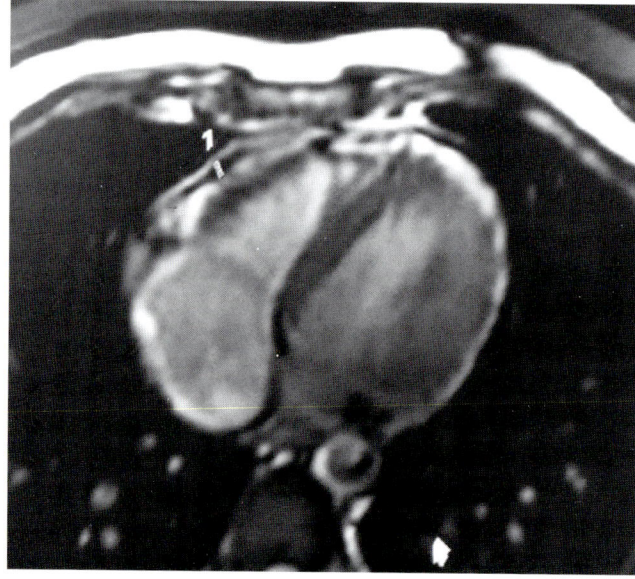

Fig. 3.9

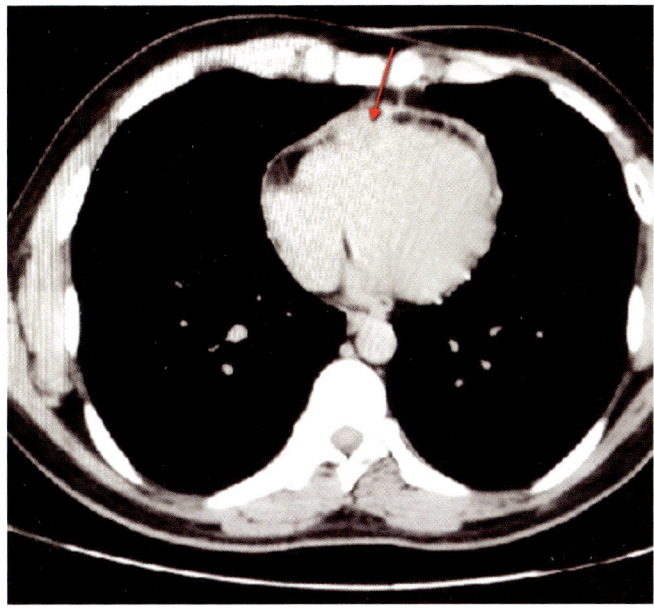

Fig. 3.10: CT scan showing lesion isodense to pericardium at free wall of right ventricle (arrow)

Surgical Intervention

Patient had ATT for 6 months and follow-up echocardiography showed resolution of the pericardial lesion.

Conclusion

The added value of CMR compared to the standard techniques used for assessment of patients with pericardial diseases has substantially increased in recent years, questioning whether this technique should not be considered the most appropriate non-invasive modality to study the pericardium. Strong points in favour of CMR are the integration of anatomic and functional information within a single examination, the ability for tissue characterization and to determine the presence and degree of inflammation and activity of disease, and the value of CMR to accurately assess the rest of the heart, in particular the myocardium, helpful in the differential diagnosis, which currently often remains a diagnostic challenge.

Suggested Reading

1. Sechtem U, Tscholakoff D, Higgins CB. MRI of the abnormal pericardium. *AJR Am J Roentgenol* 1986; 147: 245–252.
2. Shabetai R. The pericardium New York, NY: Grune and Stratton, 1981.
3. Levy-Ravetch M, Auh YH, Rubenstein WA, Whalen JP, Kazam E. CT of the pericardial recesses. *AJR Am J Roentgenol* 1985; 144: 707–714.

Case 3

Patient is 40 years old male; has breathlessness. Echocardiography was done which showed massive pericardial effusion. Pericardial thickening was mildly increased. MRI was advised to rule out constrictive pericarditis.

Cardiac MRI and Corroborative CT Scan Findings

- Massive pericardial effusion with Heterogeneous signal in fluid (Figs 3.11a and b, 3.12)
- Mildly thickened pericardium.
- Mildly enhanced pericardium on post contrast images (Fig. 3.13).
- Septal bouncing in cine sequence.
- Borderline normal LV function.
- No pericardial calcification and normal lung fields on CT scan (Figs 3.14a and b).

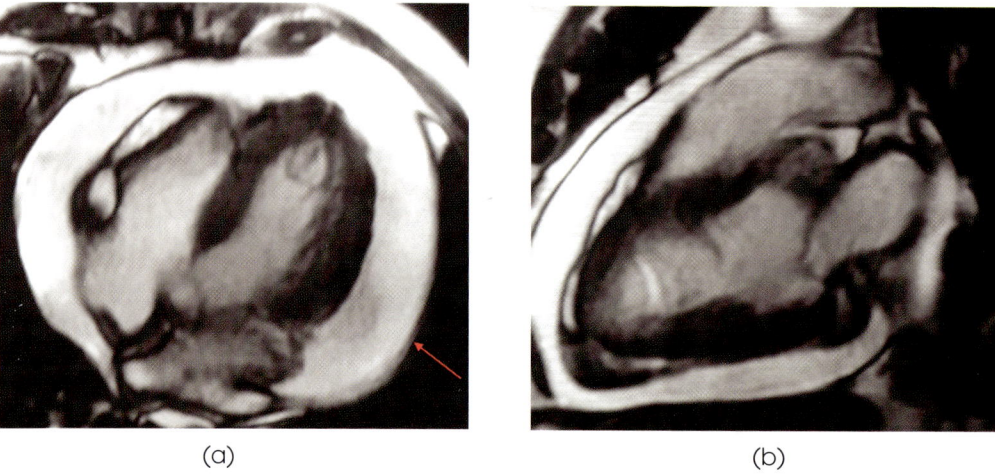

(a)   (b)

Fig. 3.11a and b: MRI showing massive pericardial effusion (arrow)

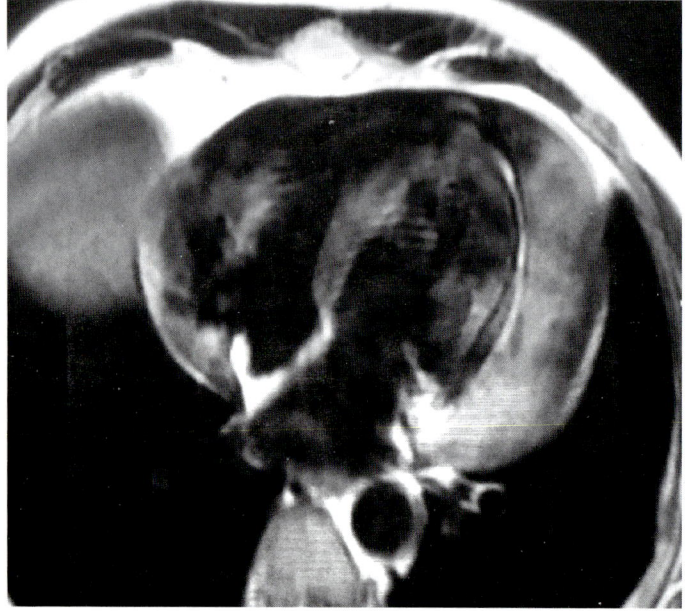

Fig. 3.12: MRI showing heterogenous signal In pericardial fluid

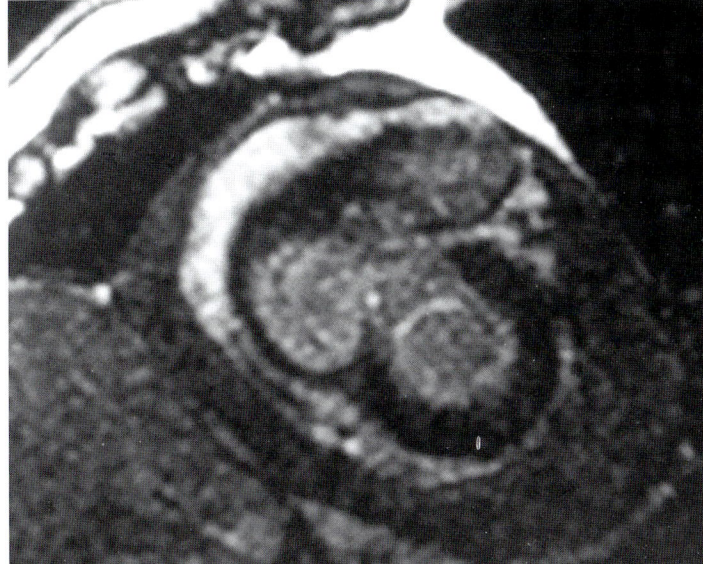

Fig. 3.13: MRI contrast image showing enhancement in pericardium

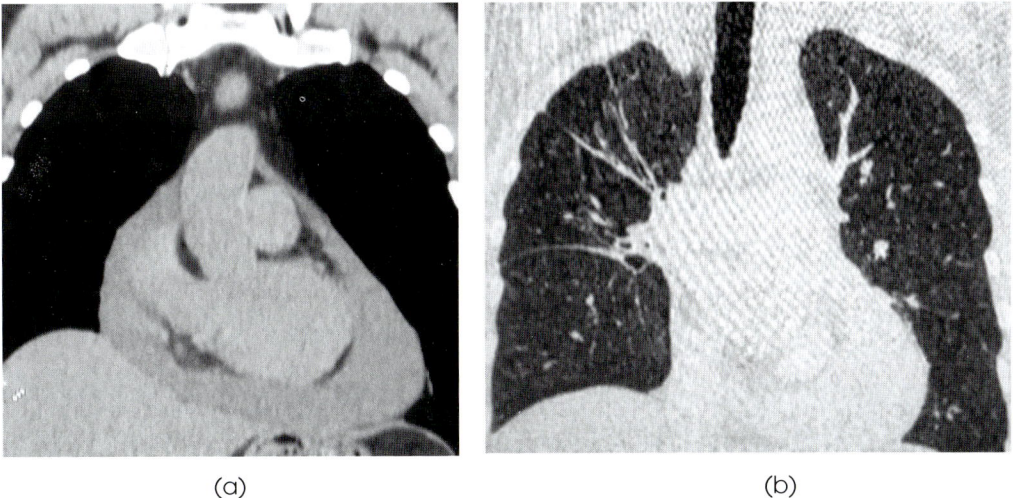

(a)   (b)

Fig. 3.14a and b: CT scan showed no pericardial calcification and normal lung fields

Radiological Diagnosis

Constrictive pericarditis and massive pericardial effusion.

Surgical Intervention

Patient underwent pericardiectomy and was discharged on ATT.

Case 4

Patient is 46 years old female came with c/o breathlessness. Echocardiography was done which showed thickened pericardium. Cardiac MRI was advised for further evaluation.

Cardiac MRI Findings

- Non-enhancing thickened pericardium measuring approx 5.5 cm (Figs 3.15 and 3.16).
- No active inflammation noted.
- Right pleural effusion is seen (Fig. 3.17)
- Normal left ventricular systolic function
- Mild biatrial enlargement
- Septal bouncing of interventricular septum
- Hypokinesia of inferior wall of left ventricle and dyskinesia of septum
- Dilated SVC and IVC.
- Tagging confirmed adherence and immobility of the pericardial-myocardial interface.
- CT showed focal calcification in pericardium (Fig. 3.18).

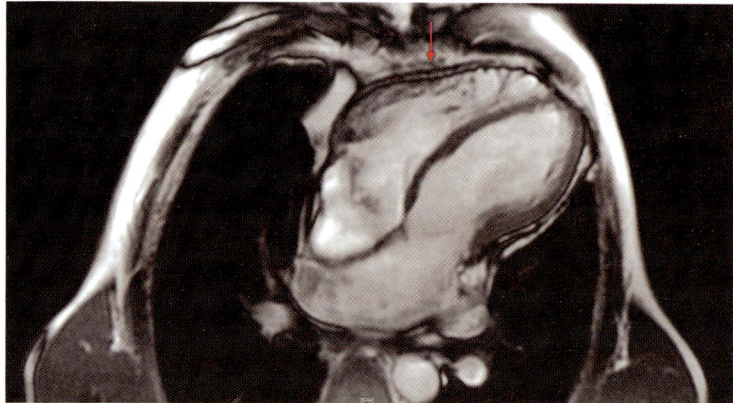

Fig. 3.15: Cardiac MRI showing thickened pericardium (arrow)

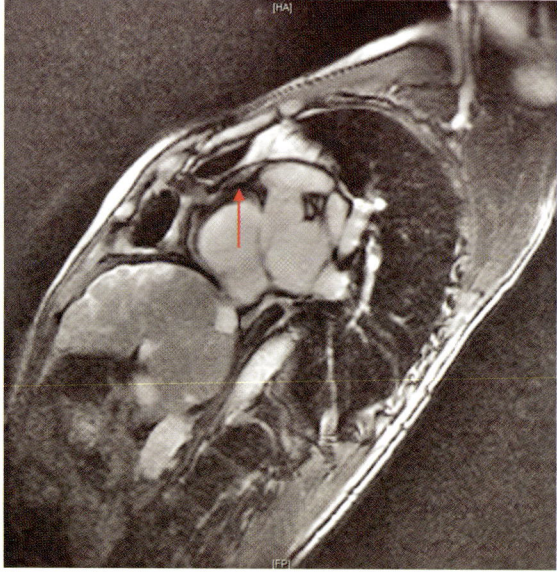

Fig. 3.16: Contrast MRI showing nonenhancing thickened pericardium (arrow)

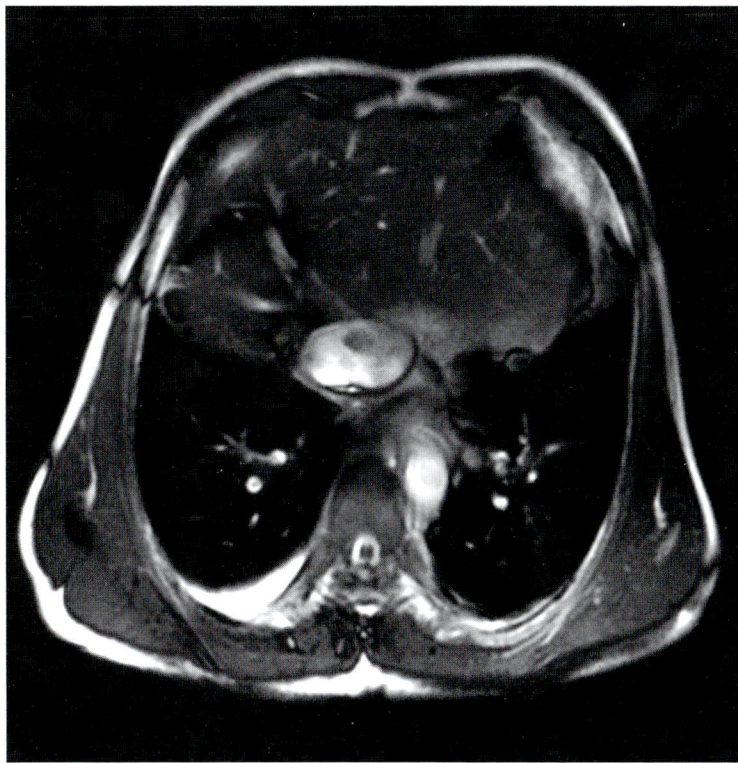

Fig. 3.17: MRI showing right pleural effusion

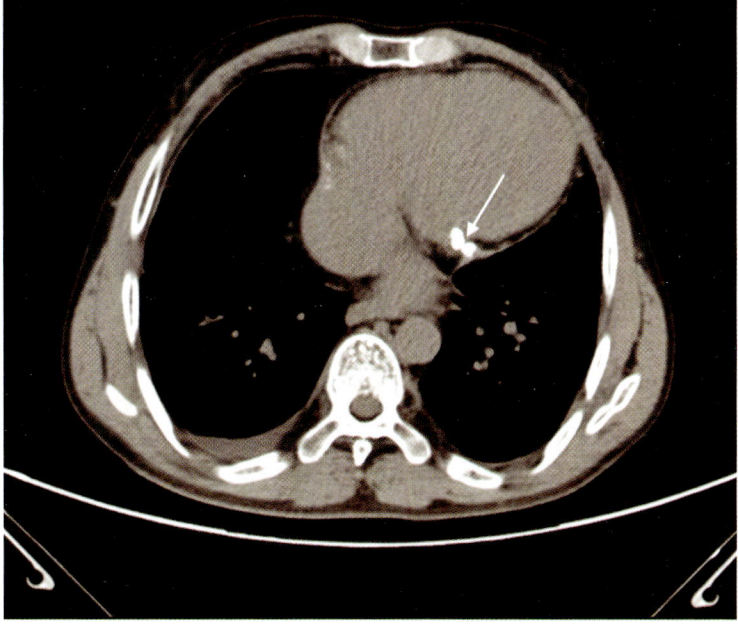

Fig. 3.18: Corroborative NCCT chest showing calcific foci in pericardium (arrow)

Radiological Diagnosis—Chronic Fibrosing Constrictive Pericarditis

Discussion: Constrictive pericarditis is a type of pericarditis which leads to diastolic dysfunction and potentially symptoms of right heart failure.

Aetiology

- Infection
- Tuberculosis: Most common cause
  - Viral infection
  - Pyogenic infection
- Rheumatic fever (rare)
- Sarcoidosis
- Previous cardiac surgery
- Radiotherapy
- Idiopathic

Types of constrictive pericarditis

- Acute—thickening with constrictive physiology.
- Effusive—chronic mainly involving visceral pericardium and associated chronic pericardial effusion.
- Adhesive—due to previous caseous pericarditis.
- Chronic fibrosing nonenhancing pericarditis.

Few MRI findings in different types of pericarditis are shown in Figs 3.19a, b, 3.20 and 3.21.

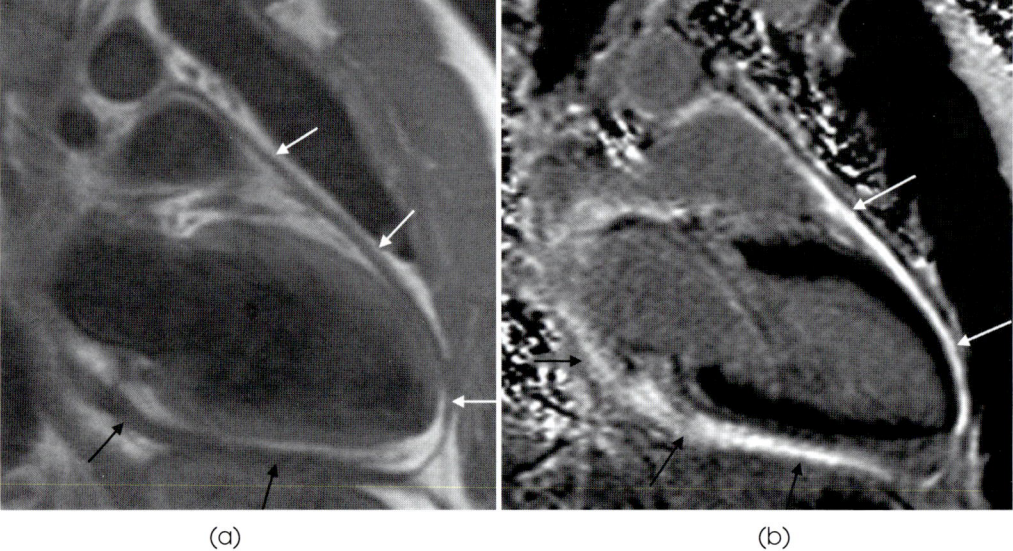

(a)   (b)

Fig. 3.19a and b: Diffuse circumferential enhancement of pericardium consistent with inflammation (arrows)

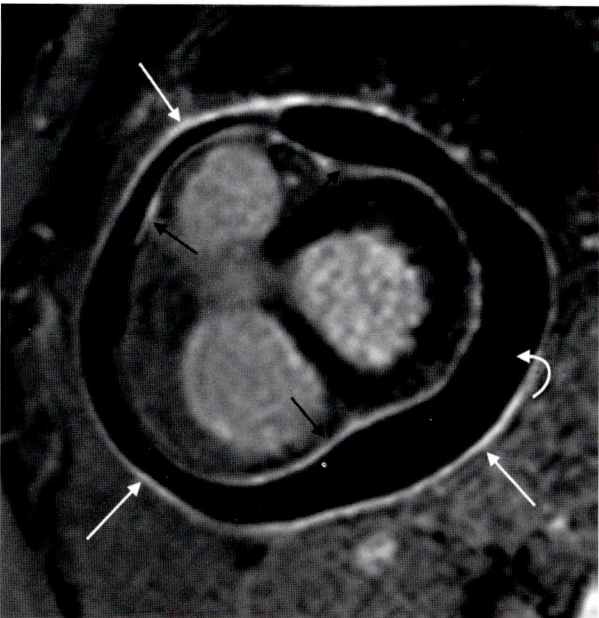

Fig. 3.20: Diffuse enhancement of thickened parietal (straight white arrows) and visceral (straight black arrows) pericardium enclosing dark pericardial effusion (curved arrow) consistent with acute pericarditis.

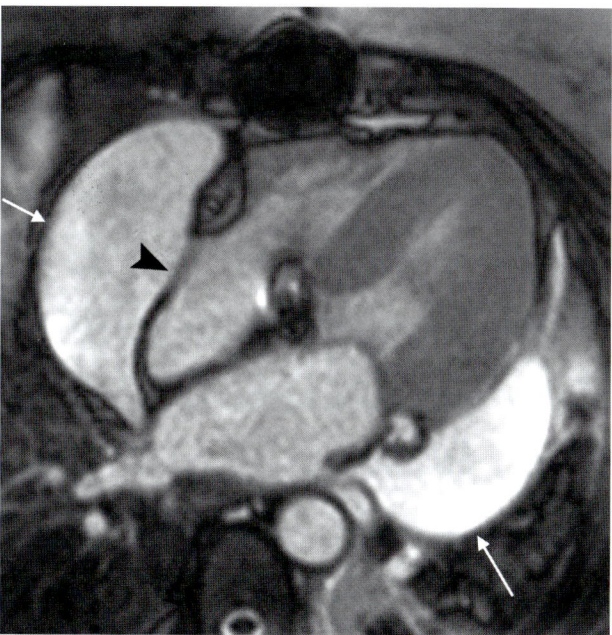

Fig. 3.21: Two loculated effusions (arrows) having high signal intensity adjacent to right atrium and left ventricle. Collapse of right atrial wall during diastole (arrowhead), which is indicative of focal cardiac tamponade.

Conclusion

- MRI has the advantages of high spatio-temporal resolution, soft-tissue contrast, wide FOV, and multiplanar imaging capabilities
- MRI is a vital diagnostic tool in the evaluation of pericardial diseases, particularly inflammation and constriction, because it can provide both morphologic and functional information essential for determining the optimal therapeutic strategy.
- MRI findings include often thickened pericardium more than 4 mm
- Septal flattening/bounce on cine MRI

Suggested Reading

1. Misselt AJ, Harris SR, Glockner J, Feng D, Syed IS, Araoz PA: MR imaging of the pericardium. Magn Reson Imaging Clin N Am. 2008, 16: 185–199. 10. 1016/j. mric. 2008. 02. 011.
2. Taylor AM, Dymarkowski S, Verbeken E, Bogaert J: Detection of pericardial inflammation with late-enhancement cardiac magnetic resonance imaging: initial results. Eur Radiol. 2006, 16: 569–574. 10.1007/s00330–005–0025–0.

Hypertrophic Cardiomyopathy

Case 1

A 29-year-old male patient visited the hospital for vague chest pain. He approached the cardiologist who advised him for an echocardiography. He had no cardiovascular risk factors, and his family history was unremarkable. His initial vital signs were stable and other physical examinations showed non-specific findings. Echocardiography showed apical hypertrophy of left ventricle. Cardiac MRI was advised to rule out if there is any infiltration in apical myocardium or any clot forming layer at apex and mimicking hypertrophy.

Cardiac MRI Findings

There was focal left ventricular hypertrophy (LVH) and wall thickness was approx 20 mm (Figs 4.1 and 4.2). Regional wall motion analysis showed hypokinesia in left ventricular apical (LVA) region. Delayed enhancement images showed no obvious pathological enhancement in left ventricular apex or in any other segment of left ventricular myocardium (Figs 4.3a and b). So infiltrative myocardial etiology was ruled out. LV qualitative analysis showed normal systolic function. LV outflow tract gradient was normal.

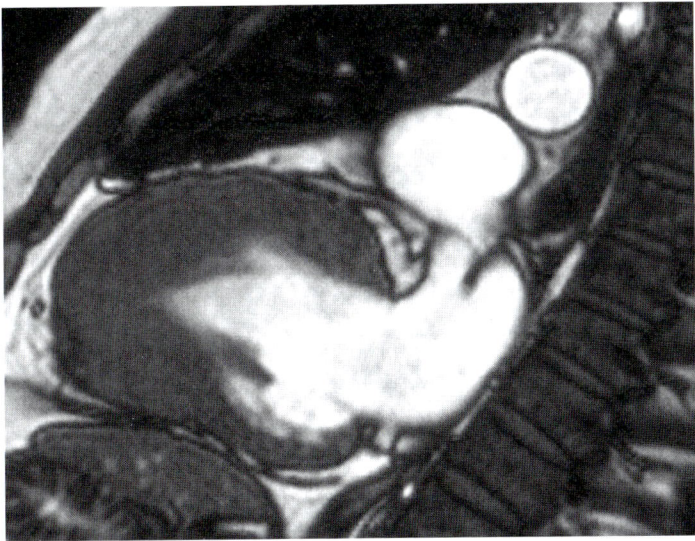

Fig. 4.1: Cardiac MRI short axis showing diffuse left ventricular apical hypertrophy

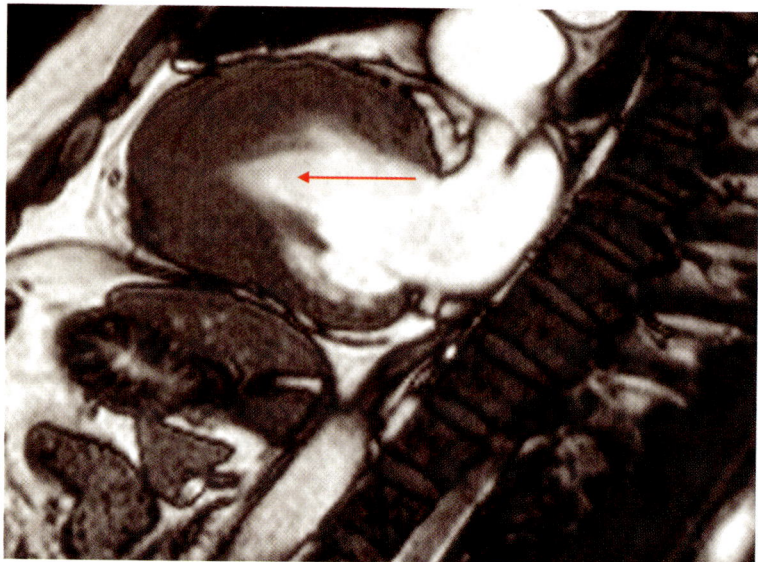

Fig. 4.2: Cardiac MRI two-chamber showing obliteration of left ventricular apical region (arrow)

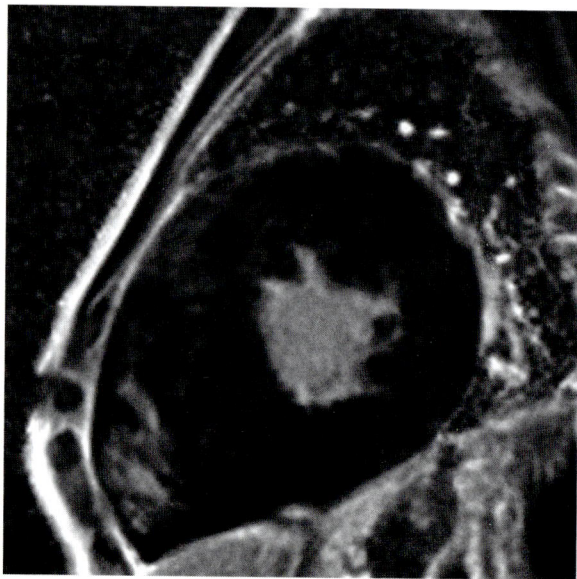

Fig. 4.3: Cardiac MRI delayed enhancement showing no pathological enhancement

Radiological Diagnosis—Hypertrophic Cardiomyopathy

Cardiac MRI was able to diagnose the case and was better in evaluating the thickness of myocardium and type of hypertrophy. Infiltrative etiology was also ruled out by MRI. Here echocardiography has a limited role.

Discussion

Cardiac MRI has the ability to evaluate wall thickness with the distribution of disease better. Also assessment of LVOT obstruction and peak outflow tract gradients can be estimated using phase-contrast MRI. Further, delayed enhancement MRI of the LV myocardium shows better infiltration if causing hypertrophy. MRI can also do the regional wall motion analysis and qualitative analysis of LV.

HCM is the most common genetic myocardial disorder and is characterized by a wide range of clinical and phenotypic expression. Non-invasive imaging is central to the diagnosis of HCM and MRI, in particular, is increasingly used to characterize abnormalities of morphology and tissue composition. In addition, MRI has an emerging role in risk stratification.

MRI screening with future longitudinal studies expected to determine which MRI features are the most important in terms of disease progression and risk of sudden death...

Suggested Reading

1. Maron BJ, Maron MS. Hypertrophic cardiomyopathy. Lancet 2013; 381: 242–55.
2. Bos JM, Towbin JA, Ackerman MJ. Diagnostic, prognostic, and therapeutic implications of genetic testing for hypertrophic cardiomyopathy. J Am Coll Cardiol 2009; 54:201.

Infective Endocarditis

Case 1

Patient is 50 years old female with history of deep vein thrombosis in 2014 with known case of papillary fibroelastoma (PF) of anterior mitral leaflet with moderate to severe MR.

Mitral valve replacement (MVR) was done on 16/9/14. Recently now complaints of orthopnoea for 3-4 days. Patient has atrial fibrillation (AF) with intermittent flutter, haemoptysis for 4 days. Echocardiography and MRI were advised.

The Echocardiography Findings

Irregularly thickened prosthetic mitral valve (Fig. 5.1).

Cardiac MRI Findings

Contrast MRI showed heterogeneous faint soft tissue enhancement surrounding the mitral valve (Figs 5.3 and 5.4). Cardiac MRI showing normal mitral valve in another patient for comparison (Fig. 5.2).

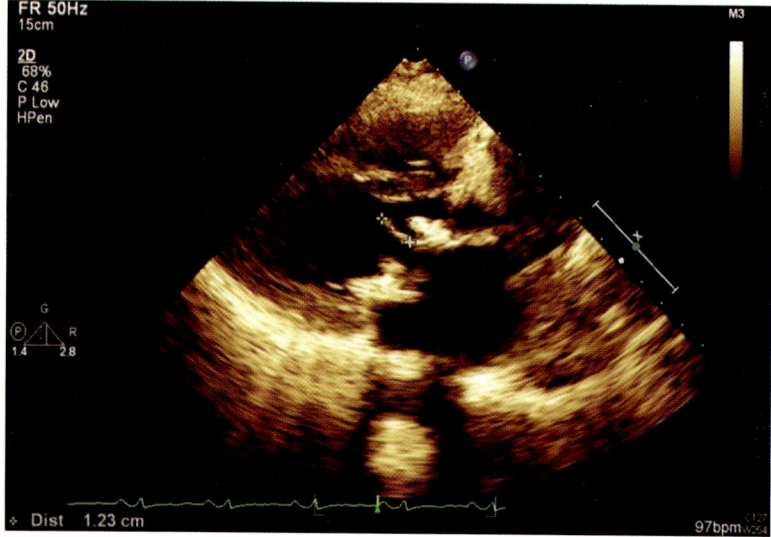

Fig. 5.1: Echocardiography showing thickened irregular prosthetic mitral valve

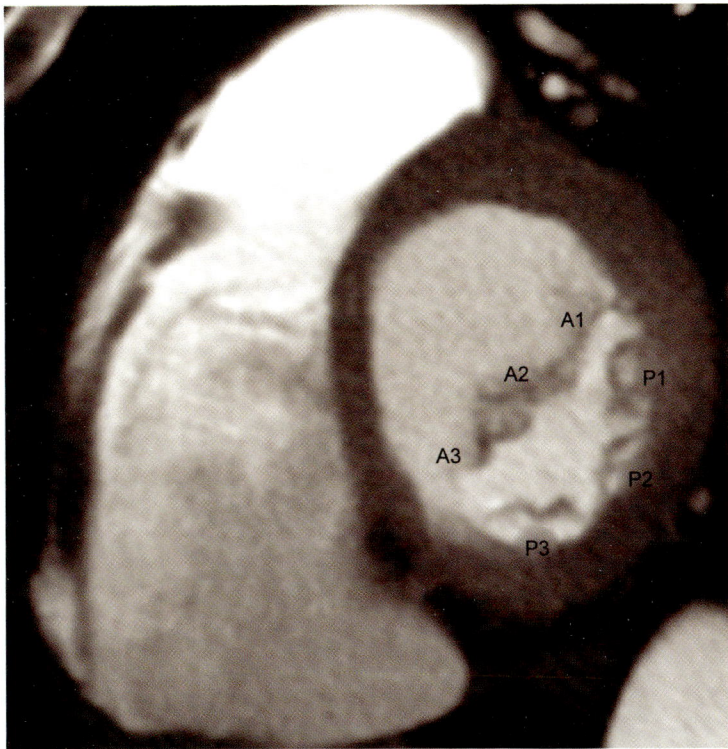

Fig. 5.2: Cardiac MRI showing normal mitral valve in another patient

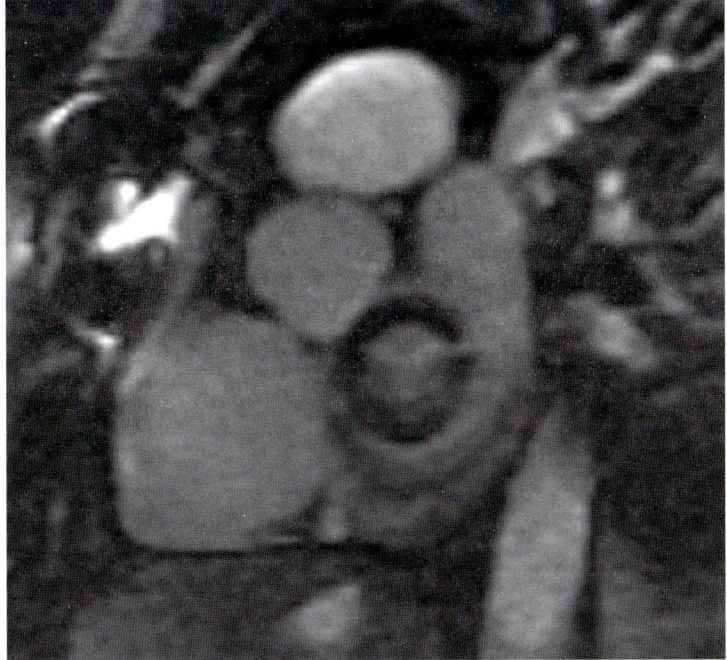

Fig. 5.3: Cardiac MRI showing thickened irregular prosthetic mitral valve in the patient

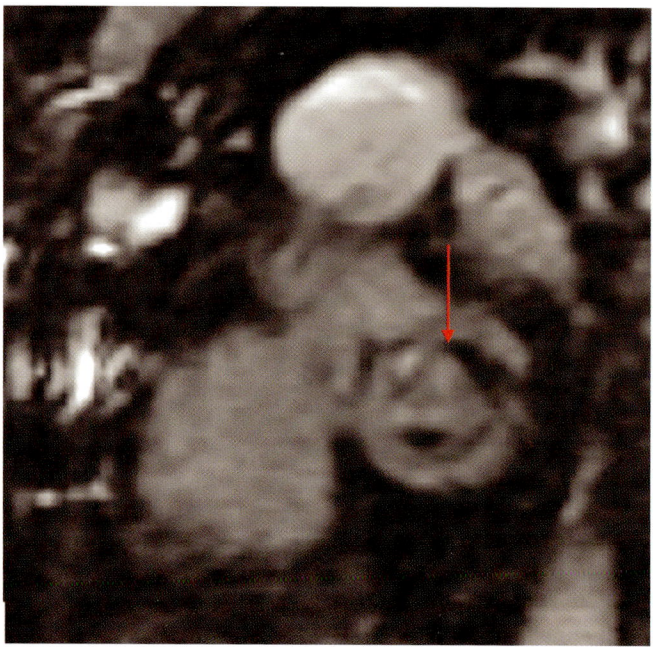

Fig. 5.4: Contrast MRI showing faint heterogenous enhancement of prosthetic mitral valve (arrow)

Radiological Diagnosis and Pathological Correlation

Cardiac MRI was done which revealed thickened prosthetic mitral valve with surrounding enhanced soft issue consistent with vegetation. 2D echo was done which revealed no signs of any valve dysfunction.

Serial blood cultures were sent in view of expected bacterial infection, and which came out positive for bacterial infection.

Treatment

Then she was being discharged on long term oral antibiotics for prophylaxis and repeat echocardiography after 6 weeks interval revealed grossly normal prosthetic valve with resolution of vegetations.

Differential diagnosis includes myxomas, thrombi, lipomas, and papillary fibroelastomas

- Myxomas usually demonstrate characteristic mobility on cine gradient-echo images. They show early moderate heterogeneous enhancement (MHE) and delayed high heterogeneous enhancement (HHE) after contrast administration (Fig. 5.8).
- Contrast-enhanced cardiac MRI reveals thrombi as low-signal-intensity, because they are avascular (Fig. 5.6).
- Lipomas demonstrate signal suppression on fat-saturated sequences (Fig. 5.7).
- Papillary fibroelastomas (PFs) appear as hypointense mobile masses on cine gradient echo images which show high signal intensity after contrast administration (Figs 5.5a and b).

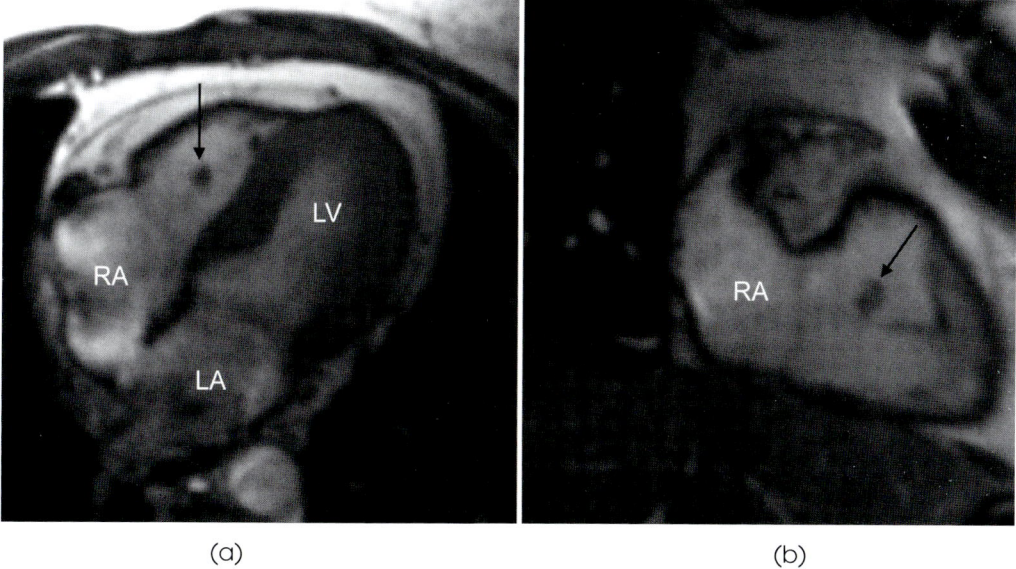

(a)   (b)

Fig. 5.5a and b: MRI showing fibroelastoma of the tricuspid sub-valvular apparatus. (a) 4-chamber SSFP image showing a well circumscribed small nodule within the body of the right ventricle (arrow). This lesion was highly mobile on cine review; (b) 2-chamber SSFP image through the right heart chambers showing the fibroelastoma attached to the valvular chordae via a thin stalk (arrow). RA; LA; LV.

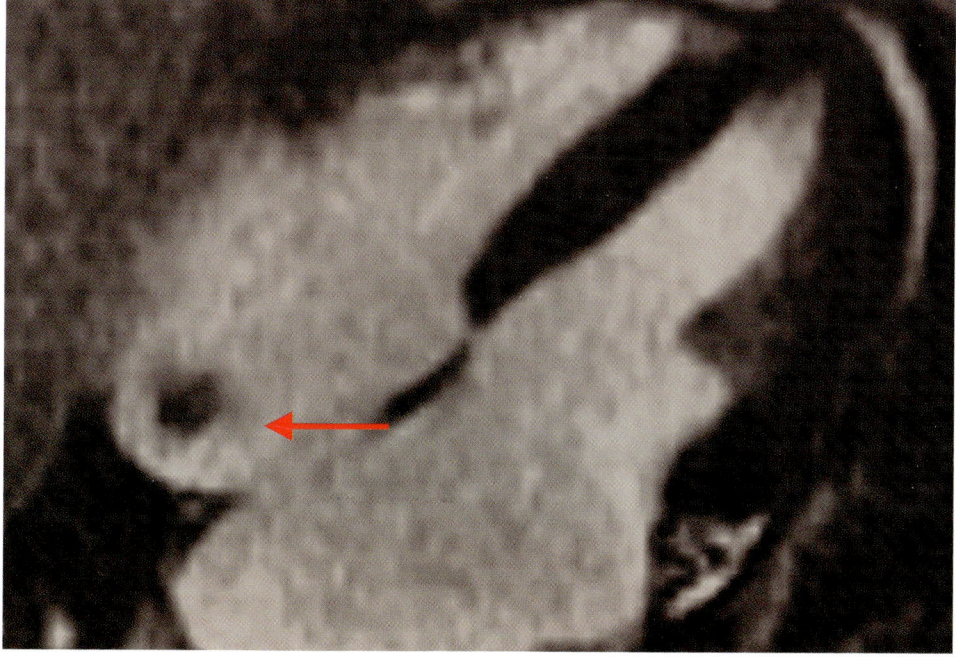

Fig. 5.6: MRI showing—RT atrial clot (arrow)

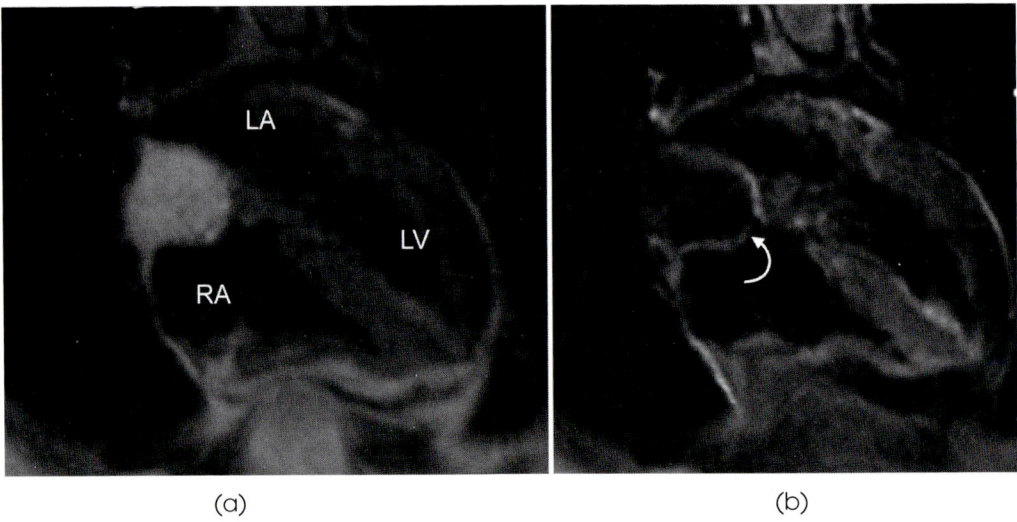

(a)   (b)

Fig. 5.7: MRI showing atrial septal lipoma. (a) 4-Chamber T2-weighted black blood image showing a well-defined uniformly high signal lesion within the Interatrial septum (curved arrow); (b) 4-Chamber T2-weighted black blood image with fat suppression showing complete and uniform signal.

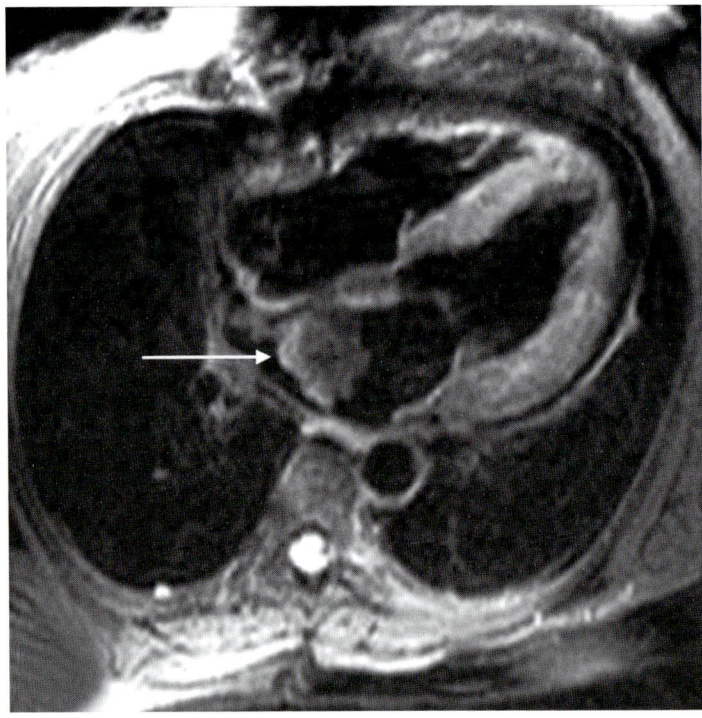

Fig. 5.8: MRI showing atrial myxoma. T1-weighted delayed phase image acquired 15 minutes postgadolinium showing patchy central foci of enhancement (arrow)

Conclusion

Cardiac MRI can be a valuable examination method to detect vegetation in patients with suspected IE.

Furthermore, MRI can give valuable diagnostic and prognostic information about the disease by depicting features such as the antegrade and retrograde dissemination, paravalvular tissue extension, and subendocardial and vascular endothelial involvement on delayed contrast-enhanced images. Cardiac MRI was also helpful in differentiating from other differential diagnosis.

Imaging of IE remains a diagnostic challenge because echocardiography has several limitations, which can impact on patient prognosis. Novel imaging modalities are emerging and offer hope of better management of the disease and thus a reduction in mortality. Some of these methods provide a better morphological evaluation of the intra- and extracardiac damage; others allow visualization of the inflammation and infection at the molecular level. Although more studies are necessary to clearly define the indications for each method, the new modalities should be included in the diagnostic criteria for IE.

Suggested Reading

1. Punja M, Mark DG, McCoy JV, Javan R, Pines JM, Brady W. Electrocardiographic manifestations of cardiac infectious-inflammatory disorders. Am J Emerg Med 2010; 28: 364–377 [Cross Ref] [Medline]
2. Skouri HN, Dec GW, Friedrich MG, Cooper LT. Noninvasive imaging in myocarditis. J Am Coll Cardiol 2006; 48: 2085–2093 [Cross Ref] [Medline]

Case 2

Patient is 14-year-old male came with history of *Plasmodium vivax* malaria in December 2015 with hypereosinophillia. He was treated for 21 days and tablet Hetrzan 2 doses were given and he also received steroids. Provisional diagnosis was made of acute leukemia, eosinophillia and malaria. Patient was referred for further management.

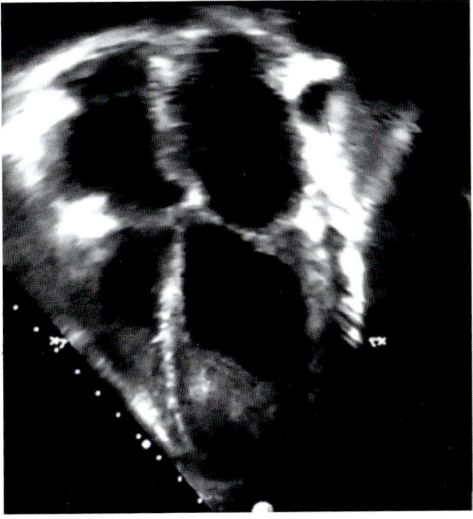

Fig. 5.9: Echocardiography showing echogenic mass in RV and LV cavity

Echocardiography was done which showed echogenic mass in left and right ventricular apical (L-RVA) region. Ejection fraction was 45%. Cardiac MRI was done for further evaluation (Fig. 5.9).

Cardiac MRI Findings

1. Dilated RA and LA size
2. Normal LV and LV systolic function
3. Subendocardial enhancement in LV and RV apex and an underlying clot (Figs 5.10a, b, 5.11 and 5.12).

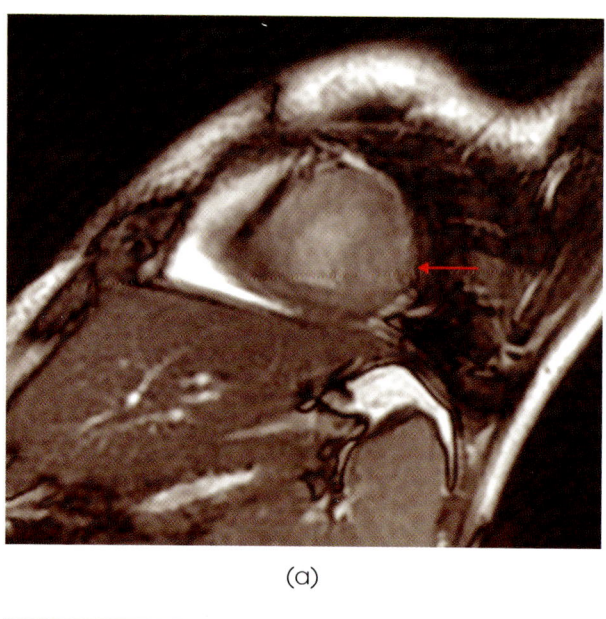

(a)

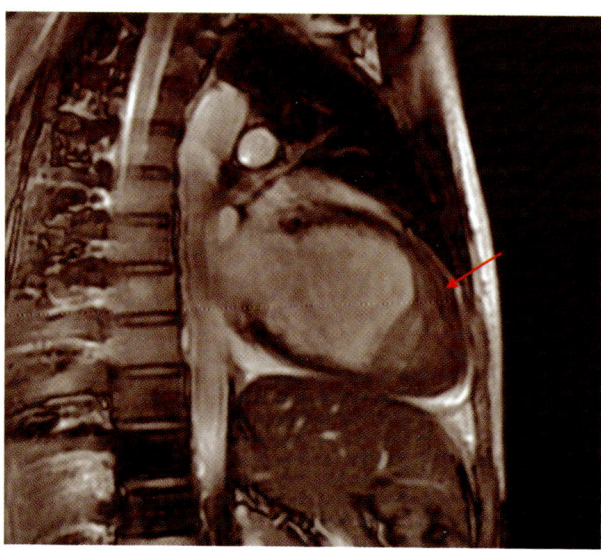

(b)

Fig. 5.10a and b: MRI showing hypointense mass in left ventricle apical region forming layer

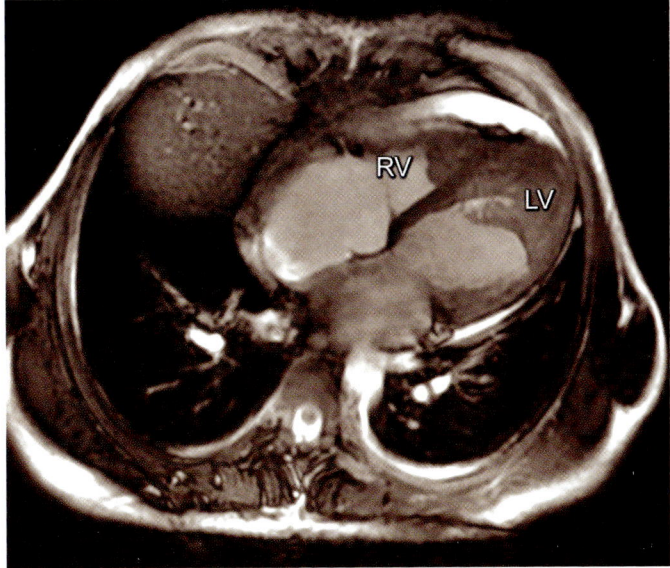

Fig. 5.11: Cardiac MRI showing hypointense mass in right and left ventricular apical region

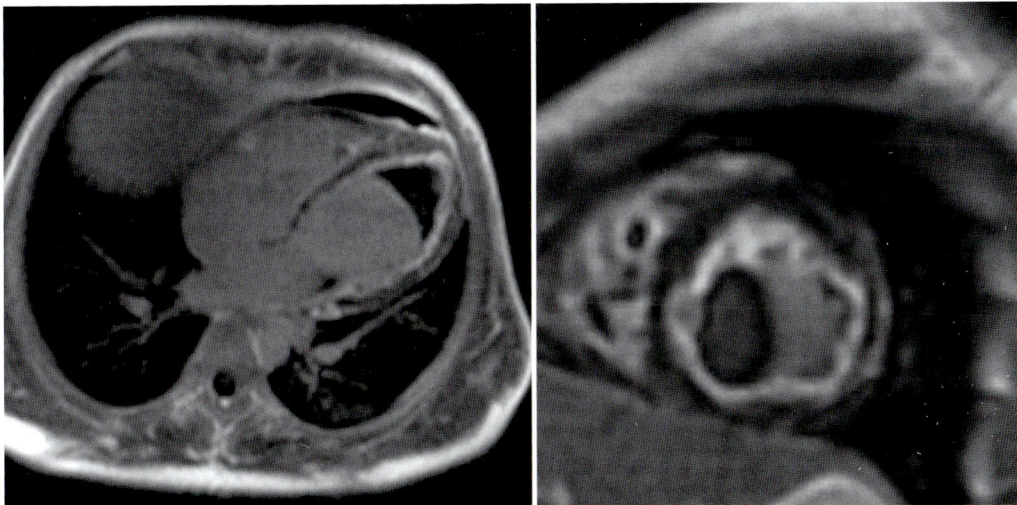

Fig. 5.12: Cardiac MRI contrast image showing subendocardial enhancement in LV and RV region

Radiological Diagnosis

Eosinophillic endocarditis.

Löffler's Syndrome (Eosinophilic Endocarditis)

- Löffler's endocarditis and cardiac manifestations of the hypereosinophilic syndrome (HES) are rare and difficult to diagnose. The transthoracic echocardiogram (TTE) demonstrated normal systolic cardiac functions and a left ventricular apical (LVA) thrombus. However, using cardiovascular magnetic resonance (CMR) with inversion-recovery (IR) delayed enhancement, and cine steady-state free precession

(SSFP) sequences, clearly demonstrate endocardial fibrosis, tissue inflammation, apical ventricular hypertrophy (VH), and LV thrombus that correlate with clinical findings.

- Epidemiologic data of end myocardial fibrosis (EMF) continues to be scarce, despite the fact that EMF is considered the most common form of restrictive cardiomyopathy (RCM).
- Previous studies reveal that about 10% of patients with rheumatoid arthritis (RA) have hyper eosinophilia (HE). They show that HE might occur either due to a reaction from the RA drug therapy or in relation to the RA process as a result of extra-articular vasculitis.

Role of Cardiac MRI

- CMR plays an important role in the diagnosis and prognosis of this condition. It provides precise morphological evaluation, usually characterized by restriction with non-dilated or small ventricles, with an element of hypertrophy, more in the apical regions.
- Among typical CMR findings, described in literature are the often increased atria size due to severe diastolic dysfunction with a restrictive disorder. IR sequences after contrast injection confirm diagnosis by showing typical subendocardial enhancement extending from the subvalvular regions to the apices of the two ventricles. Thrombi are frequently described at the apex of the LV and/or RV.

Disadvantages of MRI

- The disadvantages of cardiac MRI are its costs, which are higher than echocardiography.
- However, the emergence of new techniques and software have allowed rapid acquisitions, and in turn, a shorter scan time as well as decreased use of contrast material in some scenarios. Moreover, in today's case the cost benefit is considered relatively high, since cardiac MRI saved the costs of a performing biopsy.

Treatment

- Medical treatment included diuretics and anti-rheumatic medication, which were used to improve dyspnea.
- Initially, the patient received low molecular weight heparin (Clexane) treatment. It was withdrawn under cover of the warfarin anticoagulant.
- The patient was advised to continue the treatment and discharged from the hospital.

A follow-up MRI was recommended after one year to assess the patient's cardiac functions and the delayed enhancement, and hence the efficacy of the treatment.

Case 3

A 3-month-old male baby was brought in the hospital with history of excessive irritability.

As per transfer summary, baby was intubated in emergency for shock and ionotropes started, then baby was on dopamine + dobutamine at 20 mics/kg/min. each + milrinone at 0.4 mics/kg/min. Vitals were as follows: Heart rate: 180/min, SpO_2 = 1005. NIBP: 88/56, Chest: B/L air entry was there.

Clinical examination revealed tachycardia, hepatomegaly and no other positive finding seen.

Echocardiography was done which revealed severe LV dysfunction, no left outflow tract obstruction was seen and no structural defect was identified.

Provisional diagnosis was post-infectious dilated cardiomyopathy with hypocalcaemia.

Then chest skiagram, abdominal ultrasound and cardiac MRI were done.

Chest Skiagram Findings

Cardiomegaly (Fig. 5.13)

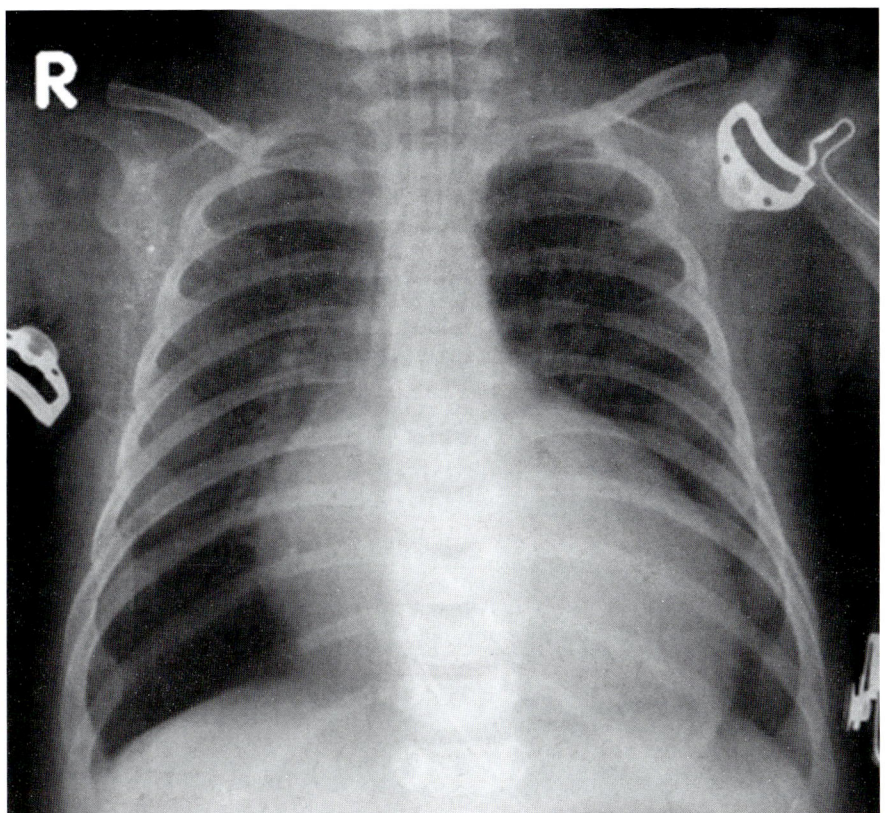

Fig. 5.13: Chest skiagram showing cardiomegaly

Ultrasonography Findings

Portal vein thrombosis (Fig. 5.14).

Cardiac MRI Findings

- Dilated LA and LV (Fig. 5.15).
- Reduced LV EF: 15%
- Diffuse endocardial enhancement in LV (Figs 5.16 and 5.17).

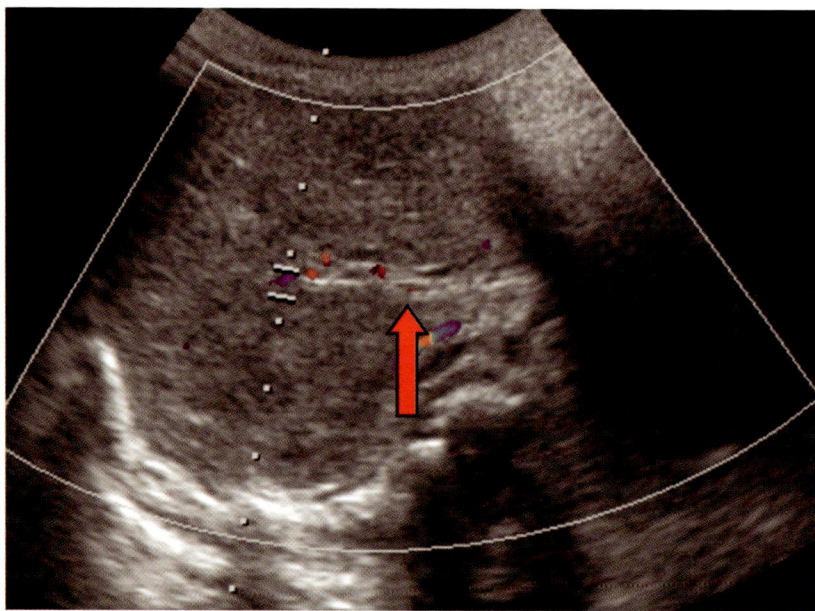

Fig. 5.14: Ultrasound abdomen showing portal vein thrombosis

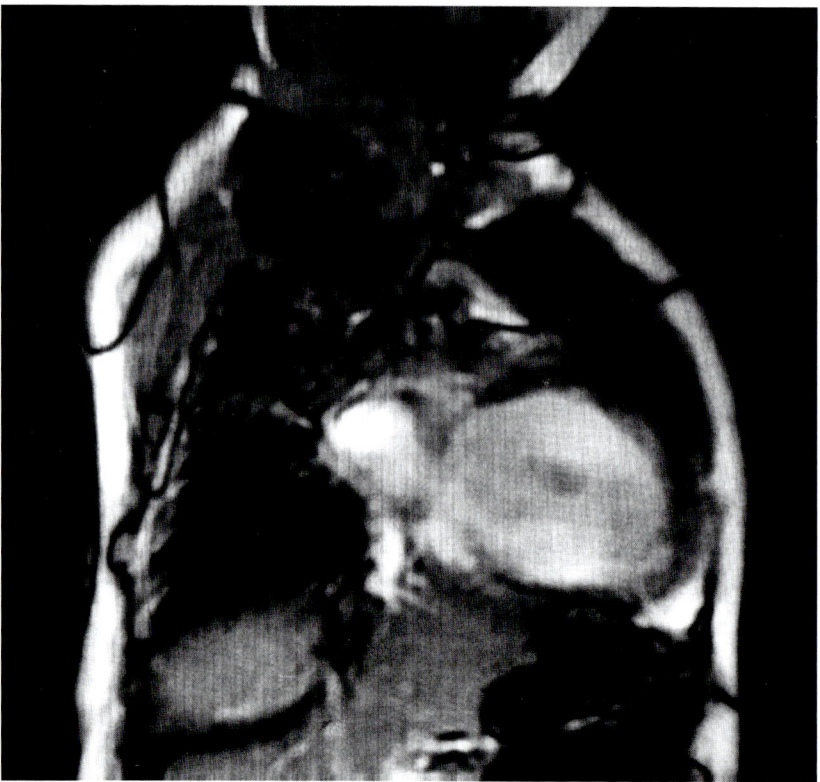

Fig. 5.15: Cardiac MRI showing dilated left ventricle

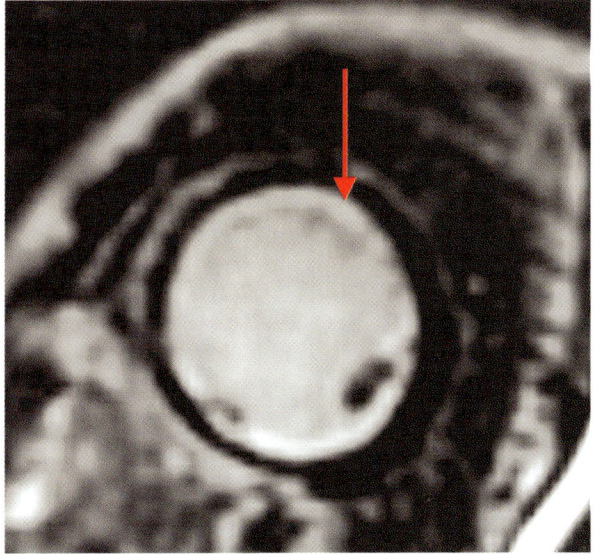

Fig. 5.16: Cardiac MRI showing subendocardial enhancement

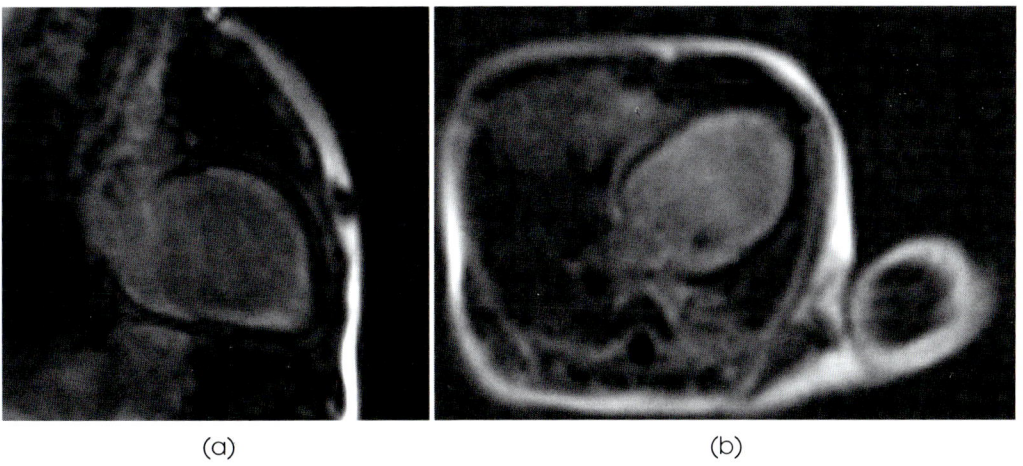

(a)   (b)

Figs 5.17a and b: Cardiac MRI showing subendocardial enhancement

Radiological Diagnosis

Endocardial Fibroelastosis (EFE)

EFE is probably a nonspecific endocardial reaction to various myocardial stress may be the result of various fetal insults such as infections, hypoxia, cardiomyopathies, immunologic diseases, vascular diseases, and chromosomal anomalies. In our case viral infection was considered.

Medical Treatment

Patient was kept on medical treatment including antiviral drugs and after an interval repeat echocardiography was done which showed improved ejection fraction.

Infiltrative Cardiomyopathy

Case 1

A 65-year-old gentleman was admitted with chief complaints of swelling in both lower limbs gradually increasing for 2 months associated with heaviness in legs. He also had shortness of breath, dizziness, fatigue and passage of frothy urine for the same duration. No history of fever, reduced urine output, haematuria, dysuria, nausea, diarrhoea, constipation, loss of weight and appetite, numbness, tingling, hypertension, diabetes, tuberculosis, asthma, and previous hospitalizations.

Histopathology—kidney biopsy was done which showed evidence of amyloidosis.

The patient was investigated. Hemogram showed normal leucocyte count raised ESR (38). RBS normal, LFT showed hypoalbuminemia. KFT was normal. Coagulation and thyroid profile was normal. Nephrotic range proteinuria (albuminuria 8.514). His chest X-ray was normal and ultrasound showed hepatomegaly with benign prostatic hypertrophy (BPH) and left kidney cyst. CT abdomen showed mild hepatomegaly with bilateral pleural effusion and diffuse abdominal wall edema.

In view of the systolic murmur, ECG and echocardiography was done. ECG showed low voltage complexes. Echocardiography was done which showed no regional wall motion abnormality of LV. Global LVEF 60%, moderate concentric LVH, increased myocardial echogenicity suggesting infiltrative cardiomyopathy—amyloidosis, LA high normal, mild MR and trace TR, Grade 1 diastolic dysfunction (E/E' >15) and minimal pericardial effusion and no intracardiac clot/vegetation (Fig. 6.1).

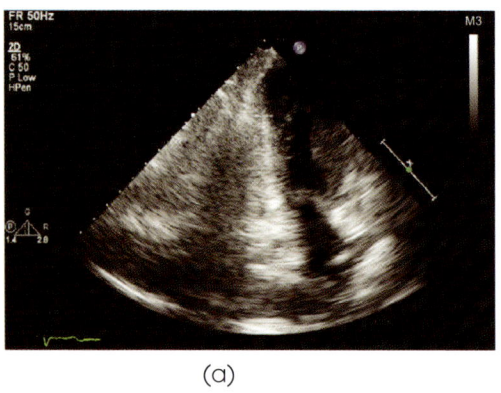 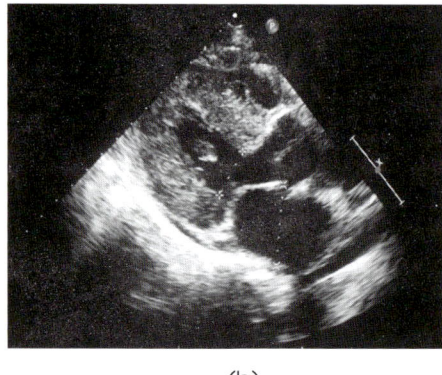

(a) (b)

Fig. 6.1a and b: Hypertrophied left ventricle with increased myocardial echogenicity (typical speckled appearance of amyloidosis) and left atria enlargement (42 mm)

Cardiac MRI showed left ventricular hypertrophy, LVEF—60%, normal LV regional wall motion, high normal LA with minimal pericardial effusion and diffuse myocardial enhancement (Fig. 6.2).

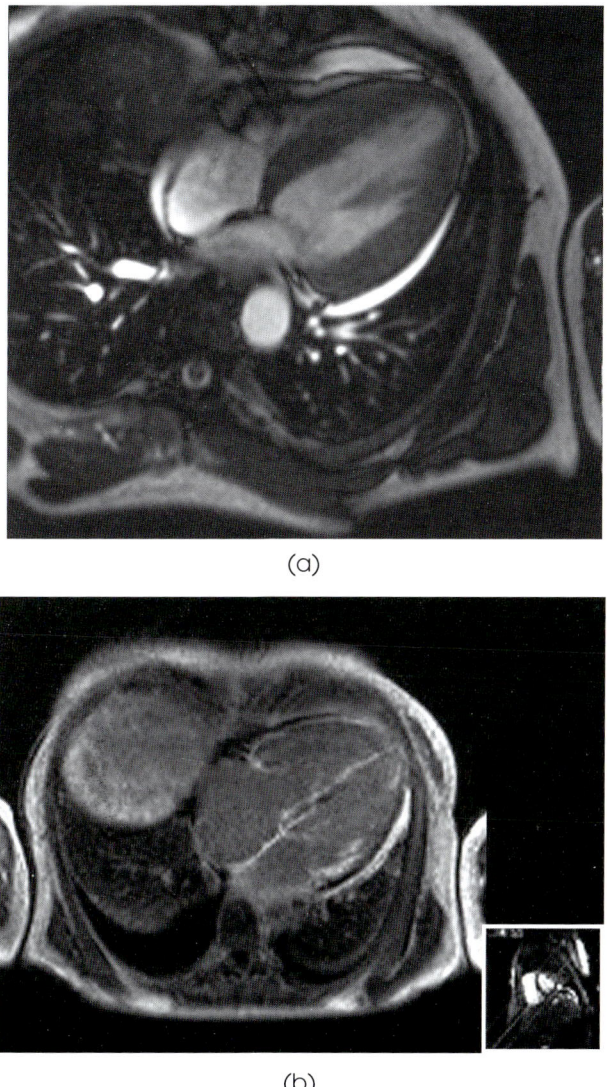

(a)

(b)

Fig. 6.2a and b: Cardiac MRI showing concentric left ventricular hypertrophy (LVH) and diffuse myocardial delayed enhancement

Discussion

We observed a number of imaging findings, the combination of which appear to be specific to amyloid cardiomyopathy. Typical features of restrictive cardiomyopathy, that is, left ventricular wall thickening, reduced systolic function with decreased ejection fraction, restriction of diastolic filling, and disproportionate atrial enlargement (AE)— were present in all cases. On delayed post-contrast images acquired 8–15 min after IV gadolinium administration, a definite widespread heterogeneous pattern of increased signal on inversion recovery T1-weighted images was observed throughout the myocardium in all of the patients. Similar changes were seen in the other patients, but they were more subtle. This pattern differs from common patterns of enhancement associated with other entities such as ischemic infarction, which usually shows intense subendocardial or transmural enhancement; infiltrative diseases such as sarcoidosis or lymphoma, in which enhancement is often focal; and interstitial fibrosis, which may show longitudinal striae of midwall enhancement. The degree of enhancement was considerably less than that seen with replacement fibrosis of myocardial infarction (MI).

Conclusion

Cardiac and renal amyloidosis is a rare finding. The presentation was unique and posed a diagnostic challenge. Prognosis of AL chain amyloidosis is poor and mortality is high. In conclusion, we consider the combination of widespread heterogeneous myocardial enhancement with other supporting features of infiltrative myocardial disease to be relatively specific for cardiac amyloidosis. Earlier diagnosis with therapeutic interventions portends a better response to current therapy and prolonged survival.

Suggested Reading

1. Kyle KA, Kobert A. Bayed, Kuwin D, and Amyloidosis: Review OP 236 Cases; Medicine: July 1975, Volume 54, Issue 4, pp. 271–299.
2. J.D. Sipe, M.D. Benson, J.N. Buxbaum Amyloid fibril protein nomenclature: 2010 recommendations from the nomenclature committee of the International Society of Amyloidosis Amyloid, 17 (2010), pp. 101–104.
3. R.H. Falk, R.L. Comenzo, M. Skinner The systemic amyloidoses N Engl J Med, 337 (1997), pp. 898–909.

Case 2

Patient is 49-year-old male, non-smoker, h/o hypothyroidism is there.

Patient has symptoms of CHF—anasarca and pedal edema.

Echocardiography was done outside which showed mild left ventricular hypertrophy (LVH), EF = 35% and global hypokinesia of left ventricular wall.

Patient was referred for cardiac MRI to rule out infiltrative cardiomyopathy.

Cardiac MRI Findings

1. Mild LVH (Fig. 6.4).
2. Delayed enhancement image showed patchy enhancement in mid-left ventricular wall (Fig. 6.5).
3. Mild hypokinesia.
4. Reduced left ventricular systolic function. EF = 31%.
5. Mild MR and TR.
6. Pleural effusion and pericardial effusion (Fig. 6.3).
7. Aortic flow analysis showed normal LVOT gradient.

O/E—Patient has a Palpable Left Supraclavicular Lymph Node

Corroborative CT scan was done which showed left supraclavicular lymphadenopathy of approximately 4.5 cm size and multiple subcentimeter to centimeter mediastinal lymphadenopathy (Figs 6.7a and b).

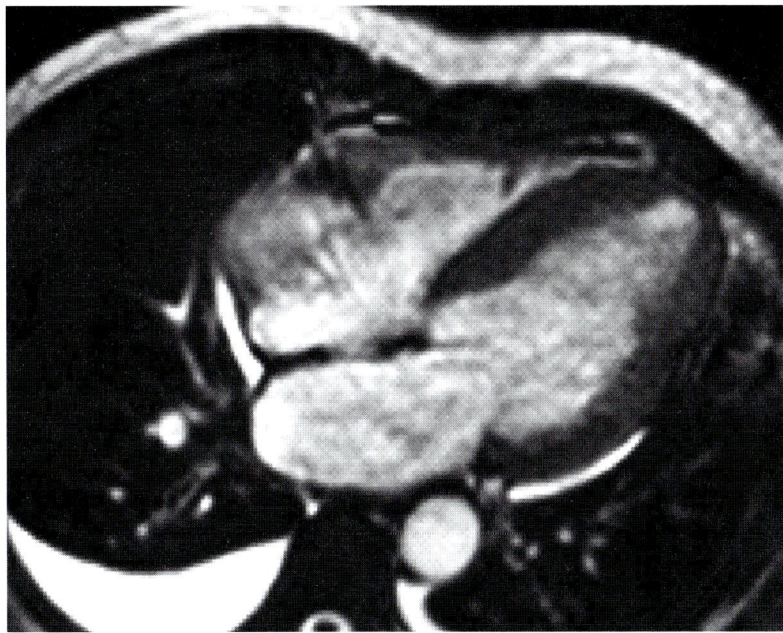

Fig. 6.3: MRI showing b/l pleural and pericardial effusion

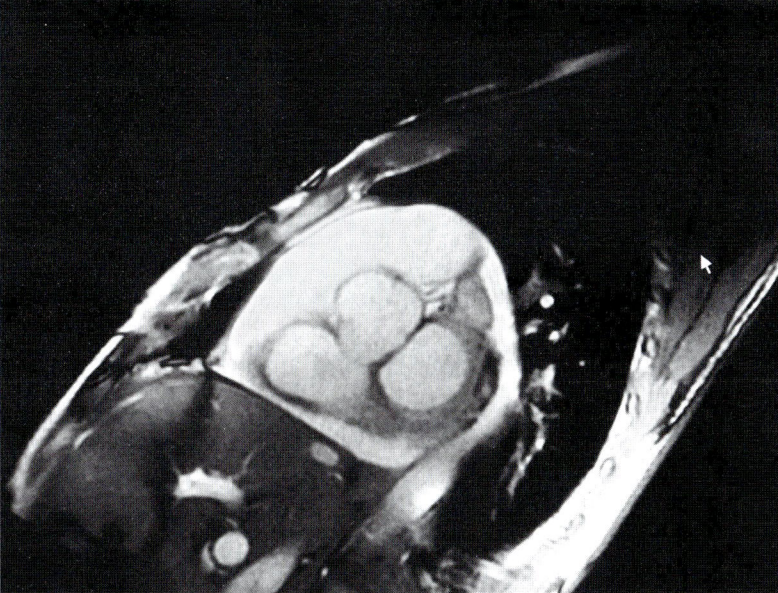

Fig. 6.4: Cardiac MRI short axis image showed mild hypertrophy of left ventricular wall

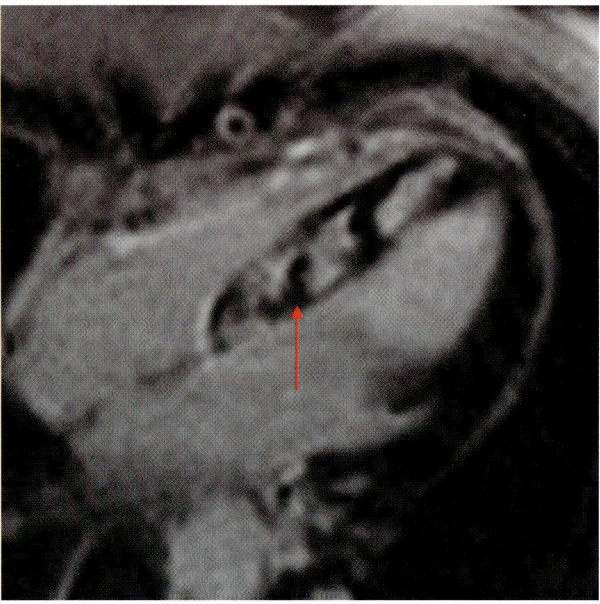

Fig. 6.5: Cardiac MRI showing patchy enhancement in left ventricular wall (arrow)

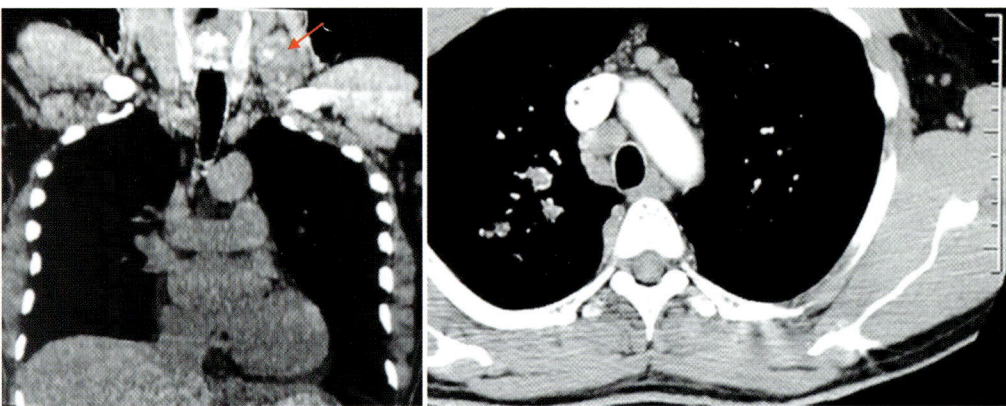

Fig. 6.6: CT scan showing left supraclavicular and mediastinal lymphadenopathy (arrow)

Differential Diagnosis

1. Sarcoidosis
2. Amyloidosis
3. Ischemic cardiomyopathy
4. Myocarditis.

Biopsy of supraclavicular lymph node and myocardium was done which came out to be sarcoidosis.

Cardiac MRI is useful in differentiating between above entities mentioned in differential diagnosis. Post-contrast myocardial enhancement is distinct in each of these.

These are as follows (Figs 6.7, 6.8 and 6.9):

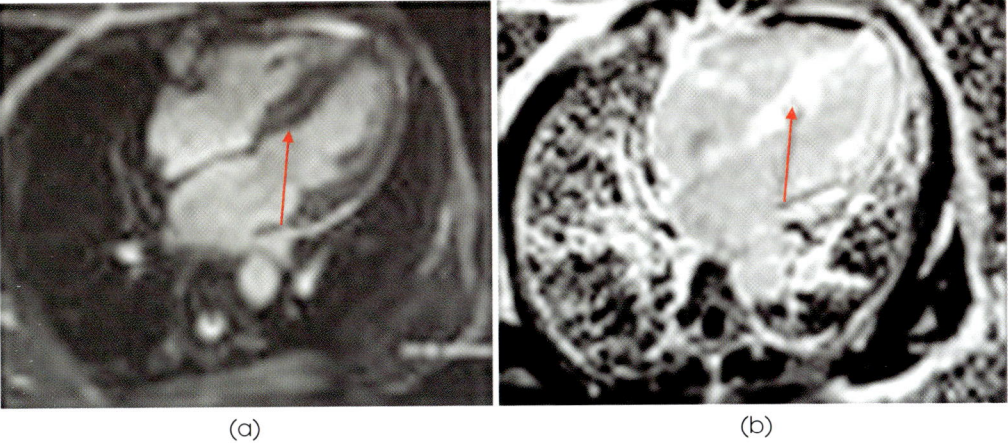

(a)   (b)

Fig. 6.7a and b: Patient with cardiac amyloidosis—late gadolinium enhancement images showed extensive amyloid infiltration (arrows) with generalized gadolinium uptake, resulting in high signal intensity and a reduced myocardial nulling time on inversion recovery sequences

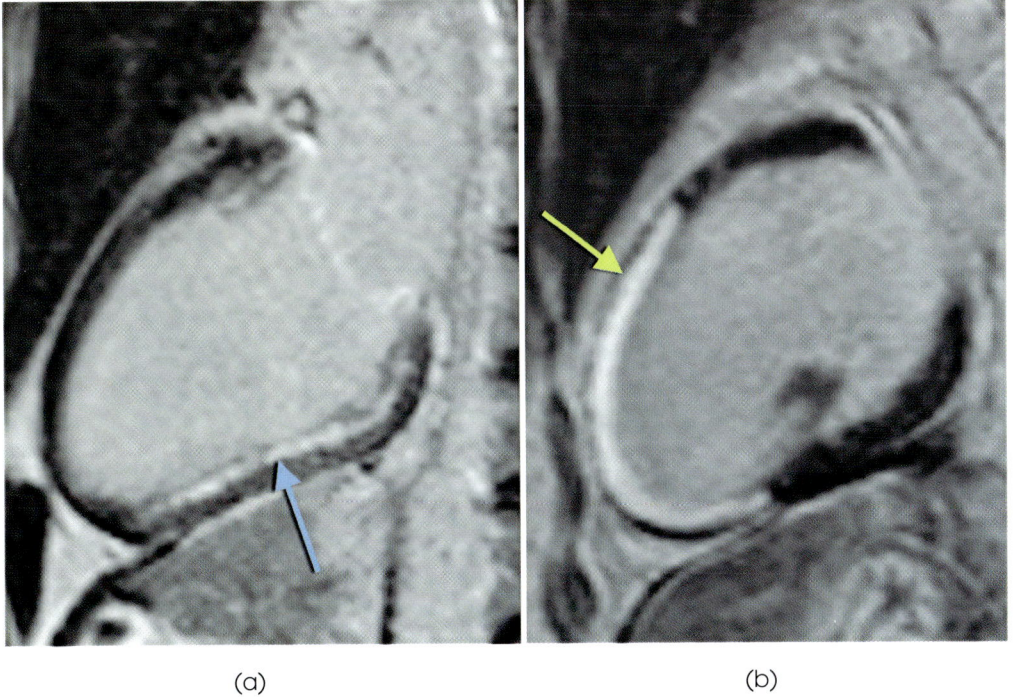

(a)   (b)

Fig. 6.8: Cardiac MRI showing subendocardial enhancement localised to vascular territory

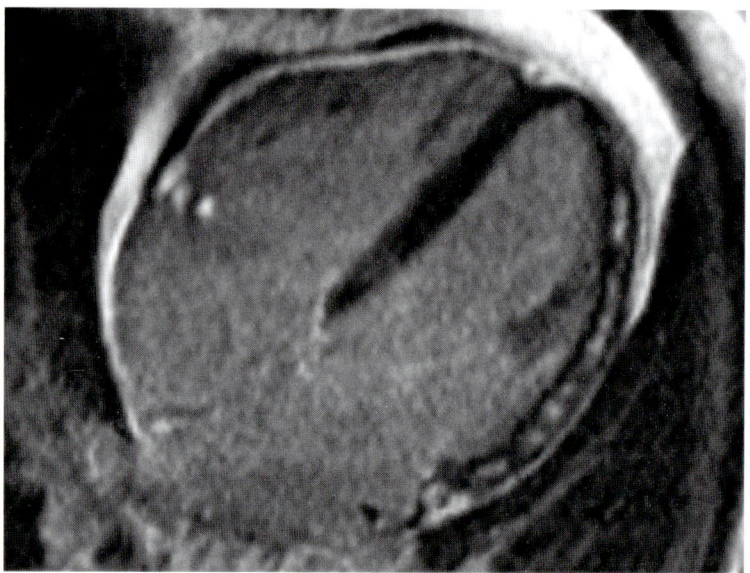

Fig. 6.9: Cardiac MRI (myocarditis) the late enhancement image showing enhancement in myocarditis is subepicardially or midmyocardially located, and does not originate from the subendocardium

Suggested Reading

1. Georgiades CS, Neyman EG, Fishman EK. Cross-sectional imaging of amyloidosis: an organ system-based approach. J Comput Assist Tomogr 2002; 26: 1035–1041.
2. Hancock EW. Differential diagnosis of restrictive cardiomyopathy and constrictive pericarditis. Heart 2001; 86: 343–349.

Intracardiac Tumour *versus* Clot

Case 1

Patient is 65-year-old, status post PTCA (no records available). Patient had a CT scan in his native city but he did not bring any report or any films. Patient had an echocardiography in the hospital which showed a large LV aneurysm arising from its posterio-lateral wall with a large clot within it (Fig. 7.1).

Patient was referred for a cardiac MRI for viability and for further evaluation. (Fig. 7.2a and b, Fig. 7.3a and b).

Cardiac MRI Findings

LV showed a large aneurysm approximately 7 cm in size arising from its posterio-lateral wall (Figs 7.2a and b).

A large clot approximately 5 cm in size is noted in the aneurysm. It showed wide-mouthed in connection with the LV. Aneurysmal wall is thin, fibrotic and shows transmural enhancement (Fig. 7.3). Rest of the myocardial segments show normal enhancement. It was showing paradoxical expansion during systole.

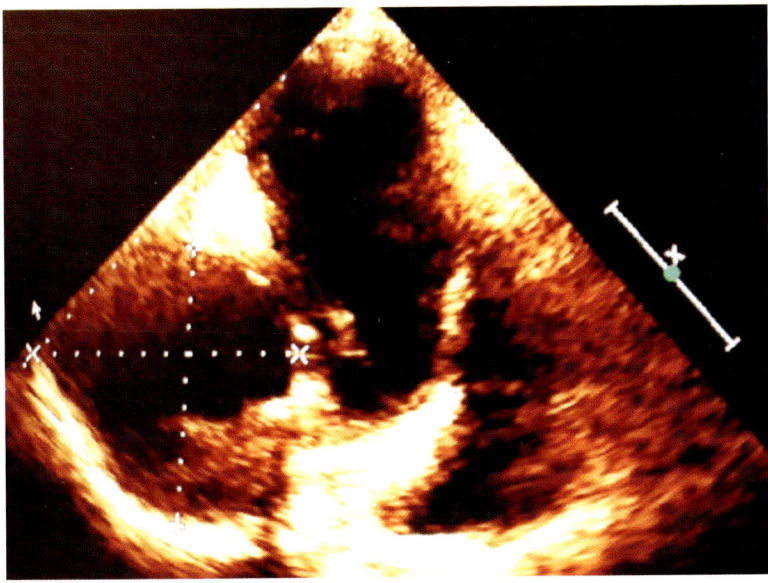

Fig. 7.1: Echocardiography in our hospital showing a large LV aneurysm arising from its posterio-lateral wall with a large clot within it

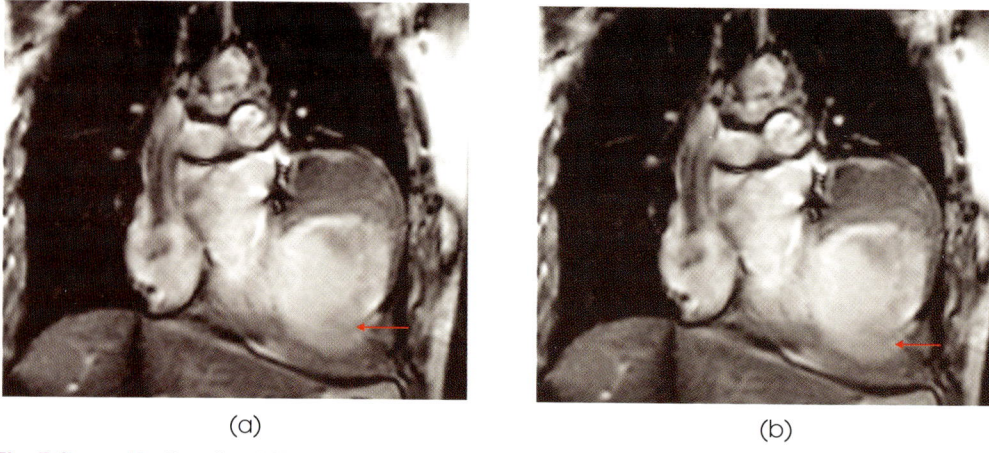

(a)   (b)

Fig. 7.2a and b: Cardiac MRI left ventricular aneurysm arising from its posterio-lateral wall (arrow)

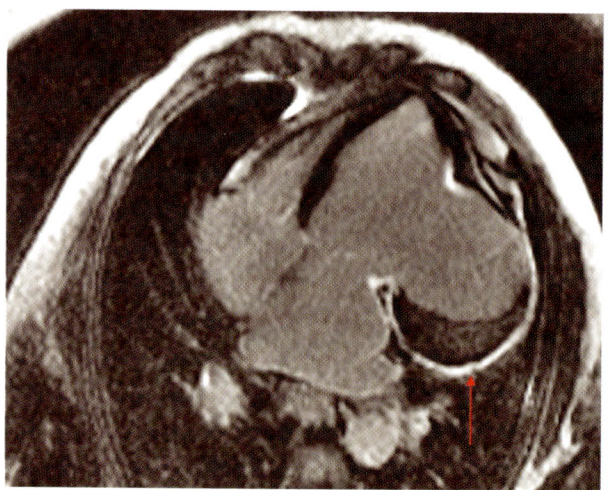

Fig. 7.3: Enhancement of the aneurysmal wall and adjacent myocardium (arrow)

Surgical Intervention

LV aneurymoraphy was done with removal of clot and stenting of circumflex artery was done as the artery was found to be 90% blocked.

Conclusion

Cardiac MRI with delayed contrast enhancement might be superior to transesophageal echocardiography (TEE) for the detection of left ventricular thrombi (LVT). This represents an additional clinically important advantage of cardiac MRI with delayed contrast enhancement for the evaluation of patients with suspected left ventricular aneurysm (LVA), and it further justifies the use of this technique in the routine evaluation of these patients. In conclusion, our findings suggest that cardiac MRI may play an important role in differentiating between true and false. True aneurysm has wide mouth connection with LV and has enhancing fibrotic wall. Further studies with delayed contrast enhancement are required to further assess the value of this feature in particular and the overall role of cardiac MRI in the diagnosis of false LVA.

Suggested Reading

1. Kumbasar B, Wu KC, Kamel IR, et al. Left ventricular true aneurysm: diagnosis of myocardial viability shown on MR imaging. AJR Am J Roentgenol 2002; 179: 472–474.
2. Gerber BL, Garot J, Bluemke DA, Wu KC, Lima JAC. Accuracy of contrast-enhanced magnetic resonance imaging in predicting improvement of regional myocardial function in patients after acute MI. Circulation 2002; 106: 1083–1089.

Case 2

A 29-year-young male came with history of stroke and sudden onset right hemiparesis. He visited a neurologist in the hospital and he was advised for a MRI brain. Brain MRI was done which showed left parietal infarct (Fig. 7.4).

Then patient was referred to cardiologist for echocardiography to rule out if any cardiac tumour or thrombus. In echocardiography there were multiple echogenic lesions in LV cavity seems to arising from apex and septal wall of LV (Fig. 7.5).

Provisional diagnosis on echocardiography was made as LV mass or clots. In view of young age with no predisposing factor like diabetes, hypertension or any h/o chest pain again cardiac tumour was considered and differential diagnosis was thrombus. To confirm this the patient was referred to radiology for cardiac MRI.

In MRI all routine sequences were done under viability protocol like axial fiesta, 2 and 4-chamber cine, short axis cine, first pass perfusion and late gadolinium enhancement (LGE).

On cine sequences LV regional function demonstrated hypokinesia of septum and apex of LV.

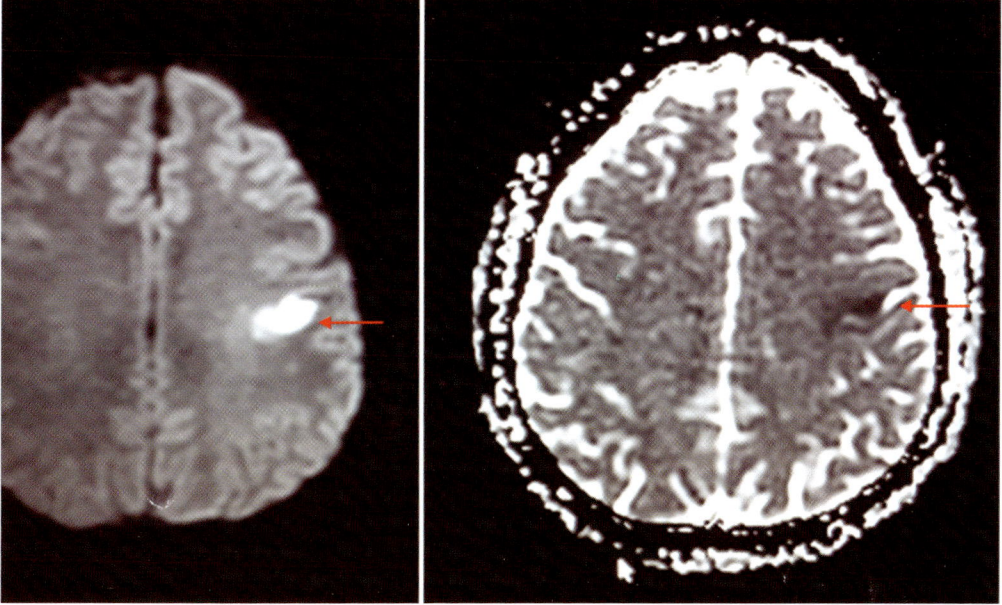

Fig. 7.4: MRI images showing left parietal infarct (arrows)

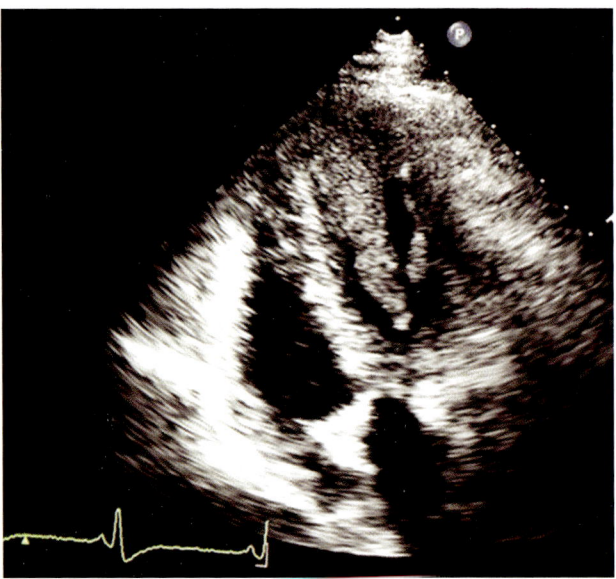

Fig. 7.5: Echogenic mass like lesions arising from apico-septal region

Short axis (SAX) and 4-chamber (4CH) images showed hypointense mass-like lesion seems to arising from apex (Figs 7.6 and 7.7).

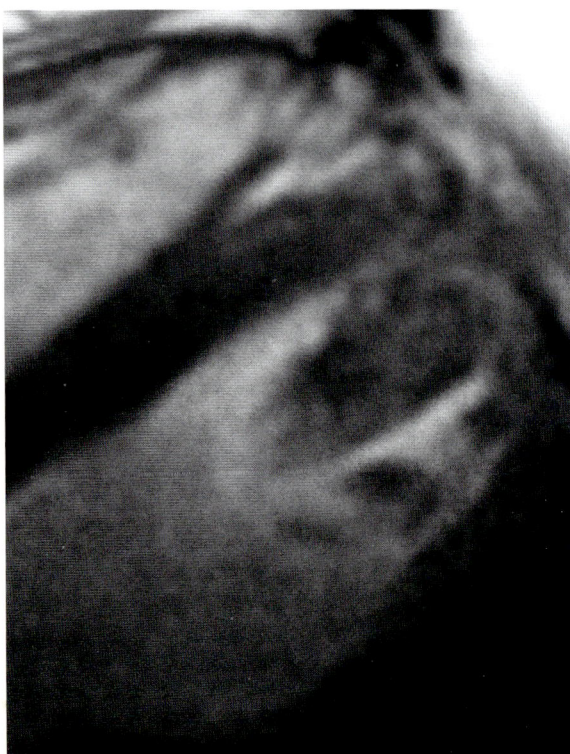

Fig. 7.6: CH image showed hypointense mass-like lesion arising from apex

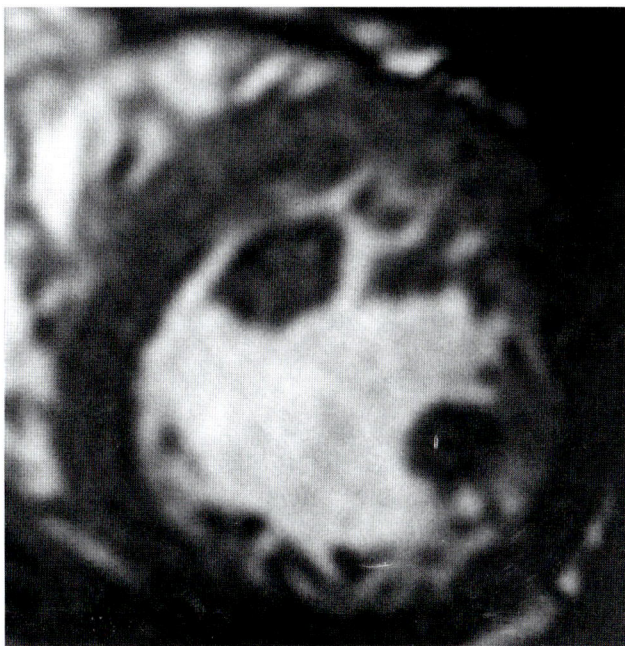

Fig. 7.7: SAX image showed hypointense mass-like lesion arising from apex

In first pass perfusion delayed arrival of the enhancement is demonstrated in anterio-septal wall of LV suggesting perfusion defect (Fig. 7.8).

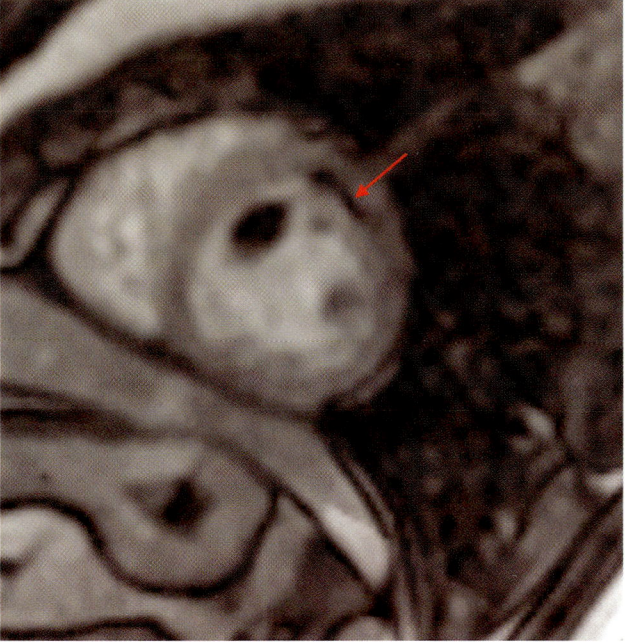

Fig. 7.8: (First pass perfusion) a SAX image showing delayed arrival of the enhancement in anterio-septal wall of left ventricle (arrow)

Then late gadolinium enhancement (LGE) image showed areas of LGE in the septal mid-apical segments (Figs 7.9 and 7.10) and in the subendocardium of the lateral and inferior apical segments.

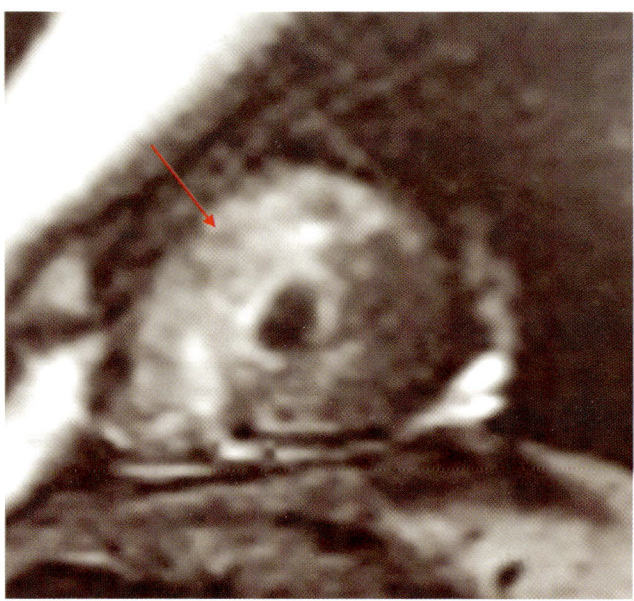

Fig. 7.9: (SAX image) showed areas of LGE in the septal mid-apical segments (arrow)

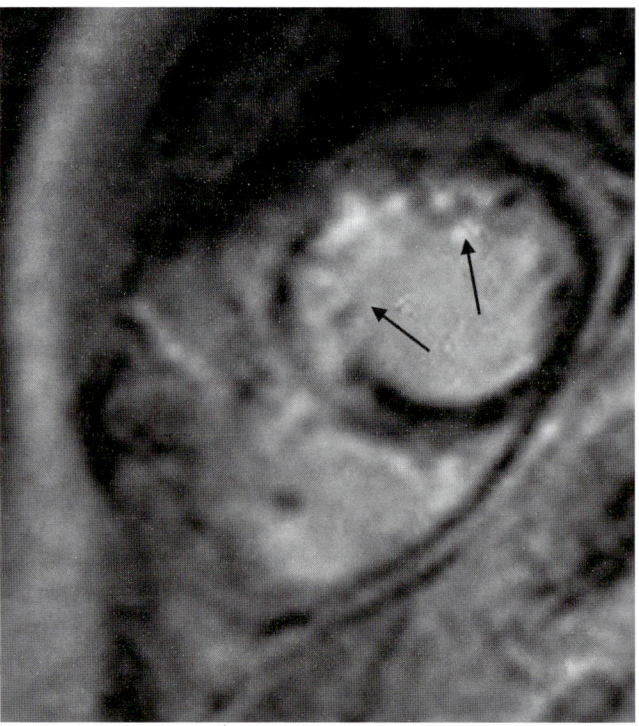

Fig. 7.10: (LGE image) showed areas of LGE in the septal mid-apical segments (arrow)

Results

MRI findings were

1. Hypokinesia of septum and apex of LV.
2. In first pass perfusion delayed arrival of the enhancement is demonstrated in anterio-septal wall of LV suggesting perfusion defect.
3. Late gadolinium enhancement (LGE) image showed areas of LGE in the septal mid-apical segments and in the subendocardium of the lateral and inferior apical segments.
4. Overall features were suggestive for myocardial infarct involving the territory of the LAD.

Conventional angiography was done which showed consistent findings.

Discussion

Cardiac MRI is a valuable modality for confirming and evaluating intracardiac thrombi. CMR identifies "mural thrombus formation" as an indication of medical necessity for CMR.

In patients with a history of ischemic heart disease or MI, ventricular thrombi frequently occur as complications. These thrombi can lead to stroke, pulmonary embolism, or peripheral arterial embolism.

Contrast-enhanced MRI provides the highest sensitivity and specificity for LV thrombus when compared to other imaging modalities, and should be considered in the care of patients at high risk of LV thrombus formation.

CMR complements other imaging modalities to detect thrombi which are hard to visualize. For example, cardiac MRI is significantly more sensitive than echocardiography for detecting ventricular thrombi (VT). Studies have demonstrated an approximately twofold increase in sensitivity for the detection of ventricular thrombi when comparison is made with echocardiography. Cardiac MRI is exquisitely sensitive for the detection of even small ventricular thrombi when certain techniques are conducted with intravenous contrast.

Suggested Reading

1. Grizzard J, Ang G. Magnetic resonance imaging of pericardial disease and cardiac masses. Cardiol clin. 2007.
2. Srichai M, et al. Clinical, imaging, and pathological characteristics of left ventricular thrombus: A comparison of contrast-enhanced magnetic resonance imaging, transthoracic echocardiography, and transesophageal echocardiography with surgical or pathological validation. American Heart Journal. 2006.
3. Barkhausen J, Hunold P, Eggebrecht H, Schuler WO, Sabin GV, Erbel R, Debatin JF. Detection and characterization of intracardiac thrombi on MR imaging. AJR Am J Roentgenol. 2002.
4. Mollet NR, Dymarkowski S, Volders W, Wathiong J, Herbots L, Rademakers FE, Bogaert J. Visualization of ventricular thrombi with contrast-enhanced magnetic resonance imaging in patients with ischemic heart disease. Circulation. 2002.
5. Barkhausen J, Hunold P, Eggebrecht H, Schuler WO, Sabin GV, Erbel R, Debatin JF. Detection and characterization of intracardiac thrombi on MR imaging. AJR Am J Roentgenol. 2002.

6. Mollet NR, Dymarkowski S, Volders W, Wathiong J, Herbots L, Rademakers FE, Bogaert J. Visualization of ventricular thrombi with contrast-enhanced magnetic resonance imaging in patients with ischemic heart disease. Circulation. 2002.

Case 3

A 29-year-old woman of Indian origin presented with history of breathlessness on exertion and fatigue for one month. General and systemic examinations were unremarkable. Routine laboratory investigations were within normal limits. Electrocardiogram was also unremarkable. Echocardiography showed a right atrial mass measuring 4.8 × 3.1 × 2.6 cm. Cardiac MRI was done for further evaluation of the mass lesion.

Cardiac MRI Findings

Well-defined hypointense lesion on all sequence is seen in right atrium seems to be abutting free wall near IVC origin (Fig. 7.11). The lesion was not enhancing. No fat component was noted in the lesion. The mass was measuring 4.8 × 3.1 × 2.6 cm. The lesion was considered as right atrial clot (Fig. 7.12).

Diagnosis—Right Atrial Clot

Well-defined hypointense lesion on all sequence was seen in right atrium seems to be abutting free wall near IVC origin. The lesion was not enhancing. No fat component was noted in the lesion. The lesion was considered as right atrial clot.

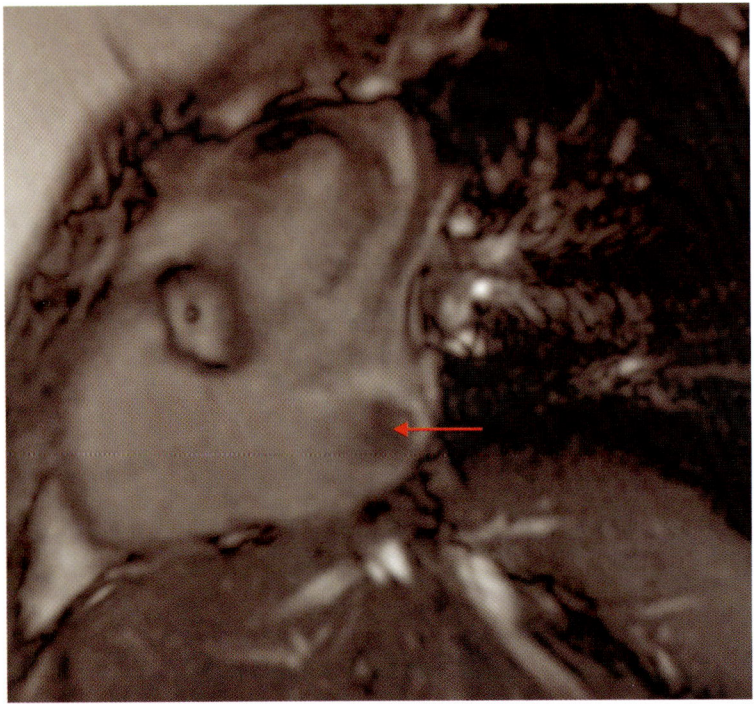

Fig. 7.11: Cardiac MRI short axis image showing hypointense lesion on all sequence in right atrium (arrow)

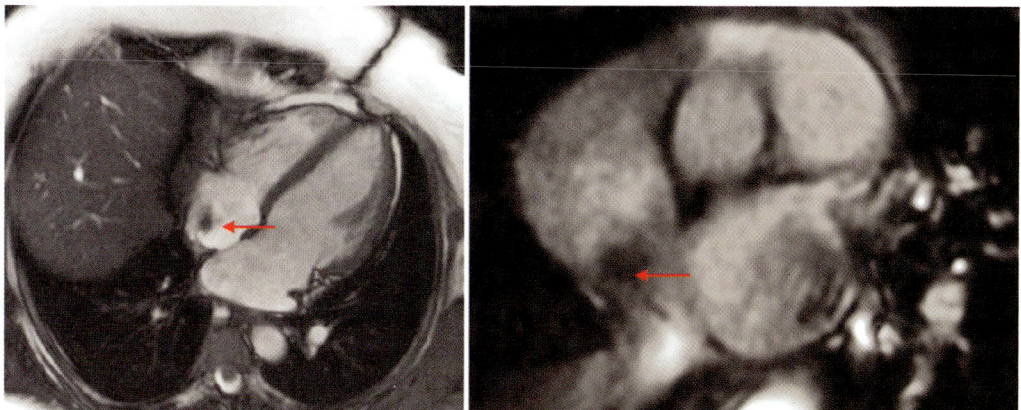

Fig. 7.12: Cardiac MRI showing hypointense lesion and no enhancement in post-contrast images the lesion is lying near IVC tip (arrows)

Surgical Intervention

The lesion was excised. Histopathological examination showed it to be organised blood clot.

Conclusions

As the prevalence of heart failure and coronary artery disease (CAD) continues to increase, the clinical importance of accurate diagnostic imaging for thrombus is heightened. Although echocardiography is widely available, it can be diagnostically limited given its reliance on the anatomic appearance of thrombus, even when image quality is judged to be optimal. DE-CMR provides tissue characterization of thrombus and has been shown to improve LV thrombus detection compared to echo-based anatomic imaging. CMR also identifies structural risk factors for LV thrombus, including infarct size/distribution and contractile dysfunction. Novel CMR techniques, including use of targeted contrast agents, may further refine thrombus characterization. Future studies are anticipated to broaden the utility of CMR in the evaluation of cardiac thrombi.

Suggested Reading

1. Shah DJ, Judd RM, Kim J. Myocardial viability. In: Edelman RR Hesselink JR, Zlatkin M, Crues JV, editors. Clinical magnetic resonance imaging. 3rd edition. New York, NY: Elsevier; 2006. pp. 1030–49.
2. Kim RJ, Fieno DS, Parrish TB, Harris K, Chen EL, Simonetti O, et al. Relationship of MRI delayed contrast enhancement to irreversible injury, infarct age, and contractile function. Circulation. 1999 Nov 9; 100(19): 1992–2002.

Case 4

A 45-year-old patient presented late after MI with massive pericardial effusion (PE) with large left ventricular apical clot. He had near total occluded Left anterior descending artery (LAD) and nonviable LAD territory. He was managed conservatively with pericardiocentesis. A cardiac MRI was advised for detailed evaluation.

Cardiac MRI Findings

Cardiac MRI was done with all routine sequences.

1. It showed thinned out scarred left ventricular apex showing aneurysmal contour and a small dehiscence in apical myocardium with underlying large apical clot (Fig. 7.13).
2. Delayed enhancement shows transmural enhancement of aneurysmal wall and adjoining myocardial segments involving LAD territory. There was a hypointense signal proceeding from the endocardium to the epicardium suggesting small dehiscence (Fig. 7.14).
3. Cine MRI images showed akinesia of left ventricular apical, mid-basal anterior wall.

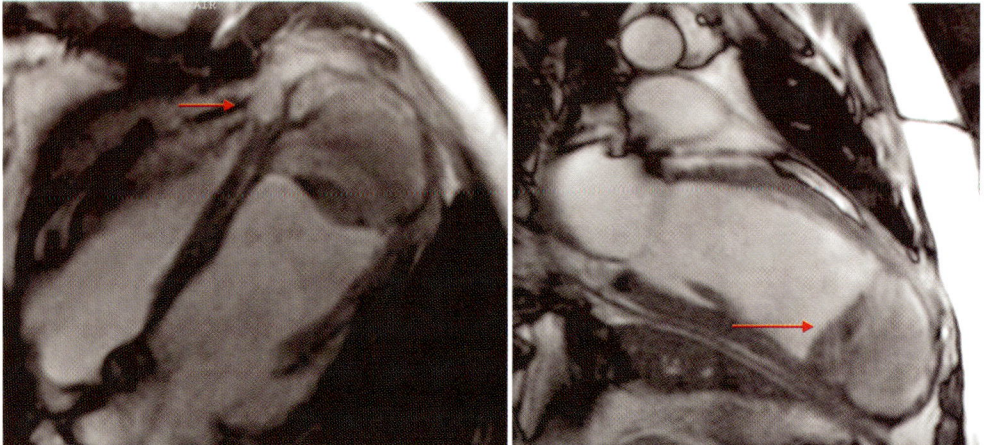

Fig. 7.13: MRI showing thinned out scarred left ventricular apex with aneurysmal contour and a small dehiscence in apical myocardium with underlying large apical clot (arrows)

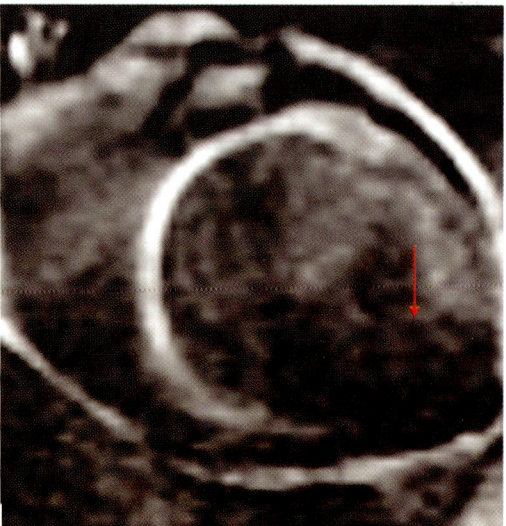

Fig. 7.14: MRI showing delayed enhancement of aneurysmal wall and adjoining myocardial segments involving LAD territory and a hypointense signal proceeding from the endocardium to the epicardium suggesting small dehiscence (arrow)

Diagnosis

Cardiac MRI showed thinned out scarred left ventricular (LV) apex with aneurysmal contour, having a small dehiscence with underlying large apical clot.

Surgical Intervention

Patient underwent aneurysmorrhaphy and grafting to LAD artery.

Conclusion

This case illustrates the importance of multimodality imaging in the clinical detection of a left venticular thrombi (LVT) when a high degree of suspicion exists. TTE alone was unable to adequately visualize the source of the filling defect demonstrated by left ventriculography. The use of CMR was necessary in this report to confirm the presence of a thrombus, serving to highlight its superior sensitivity and specificity, as well as the limitations of other imaging modalities in thrombus detection.

Suggested Reading

Srichai MB, Junor C, Rodriguez LL, Stillman AE, Grimm RA, Lieber ML, et al. Clinical, imaging, and pathological characteristics of LVT: A comparison of contrast-enhanced MRI, transthoracic echocardiography (TTE), and transesophageal echocardiography (TEE) with surgical or pathological validation. Am Heart J. 2006; 152: 75–84.

Case 5

Patient came with complaints of breathlessness off and on. He is known case of HTN, type II DM taking treatment.

Echocardiography was done which showed heterogeneous echotexture mass in left atrium attached to interatrial septum. The differential diagnosis on echo was atrial myxoma and clot (Fig. 7.15a and b).

Cardiac MRI Findings

Heterosignal signal intensity mass on T_2-weighted sequence was seen in left atrium attached to interatrial septum. The mass was measuring 3.7 × 2.7 cm in size, relatively

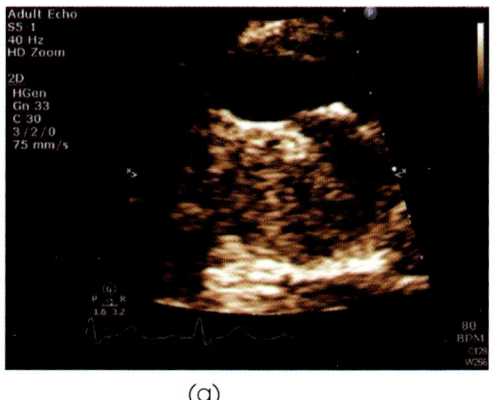 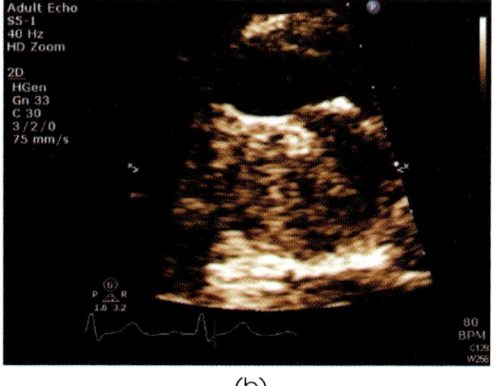

(a)　　　　　　　　　　　　　　　　　(b)

Fig. 7.15a and b: Echocardiography showing left atrial mass attached to inter-atrial septum

mobile and showing heterogeneous delayed enhancement. The mass had a pedicle by which it was attached to the septum. Rest of the cardiac chambers were normal in dimension (Figs 7.16 to 7.18). LV functions were normal. Regional wall motion analysis was showing normal LV atrial wall motion.

Corroborative noncontrast CT scan was done just after MRI which showed that contrast which was given during MRI was retained in the mass (Fig. 7.19).

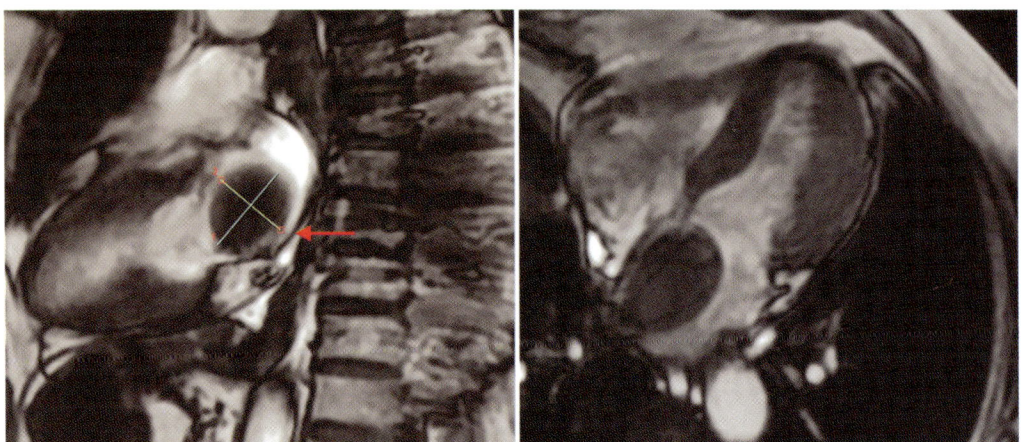

Fig. 7.16: Cardiac MRI 2 and 4 chamber image showing mass in left atrium and attached to inter-atrial septum (arrow)

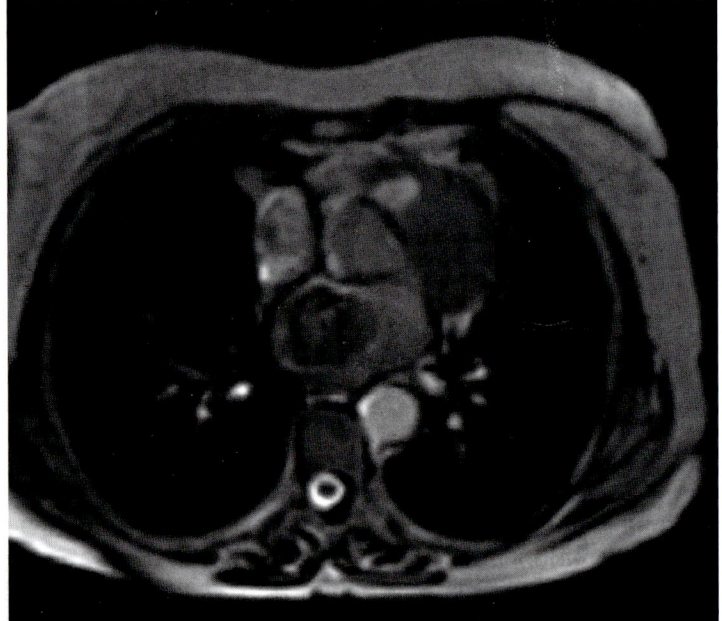

Fig. 7.17: Cardiac MRI T2 weighted sequence showing heterogenous signal intensity of the mass.

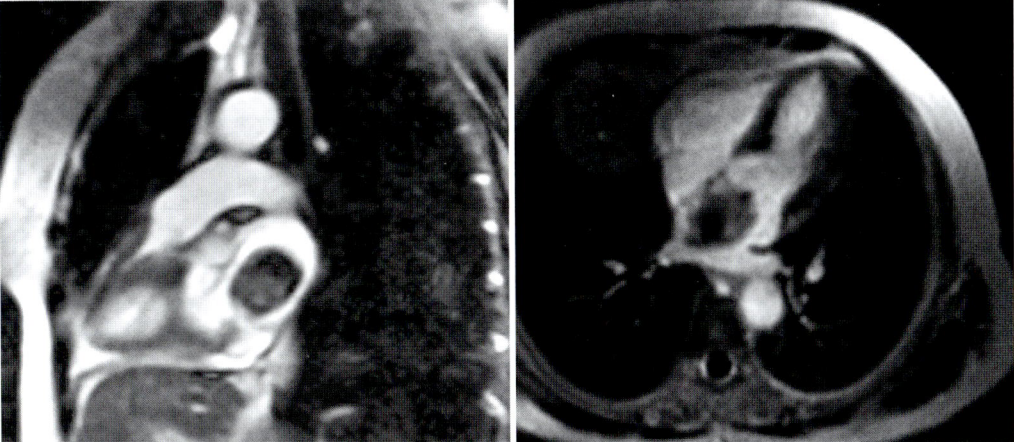

Fig. 7.18: MRI delayed enhancement image showing heterogenous enhancement of the mass

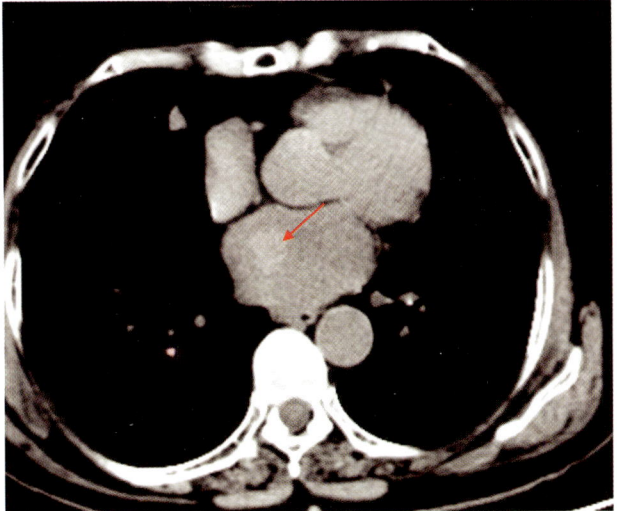

Fig. 7.19: Corroborative CT scan showing MRI contrast retained in the mass and the scan was done just after the completion of MRI (arrow)

Surgical Intervention

Excision of left atrial mass was done. Biopsy confirmed it to be atrial myxoma.

MR is helpful in differentiating between mass and clot as thrombus is T1 hypo-intense (darker) as compared to normal muscle and usually does not enhance except chronic thrombus. Mass usually shows mild to moderate enhancement and have different tissue characterization as compared to normal myocardium.

Case 6

Patient is 58 years old female came with complaints of fever, weakness and loss of weight. Patient was referred for contrast CT chest.

Contrast CT Chest Findings (Figs 7.20a, b and 7.21a, b)

Large hypodense mass of 5.2 × 4.6 cm size is seen in right atrium causing bulging of tricuspid valve into RV. It shows thin peripheral calcification with focal delayed enhancement (30 mins) in the mass.

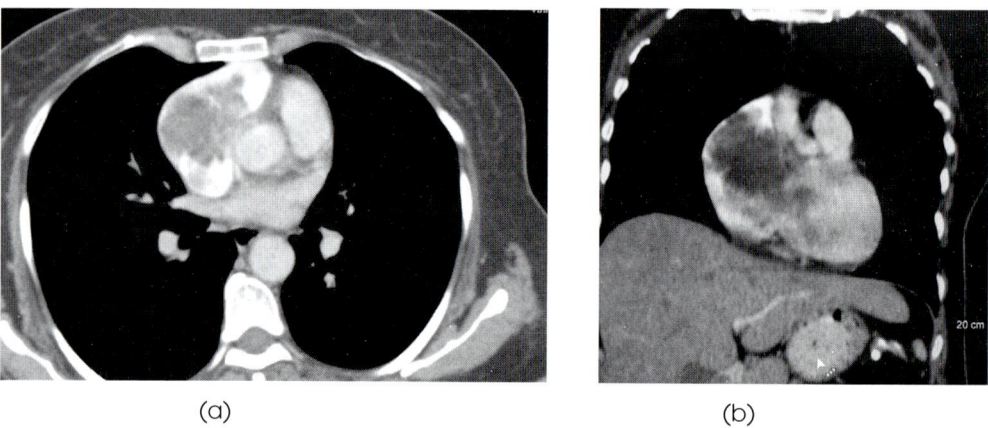

(a)　　　　　　　　　　　　　　　　　　　(b)

Fig. 7.20a and b: CECT chest showing hypodense mass lesion in right atrial cavity causing bulging of the tricuspid valve into right ventricle

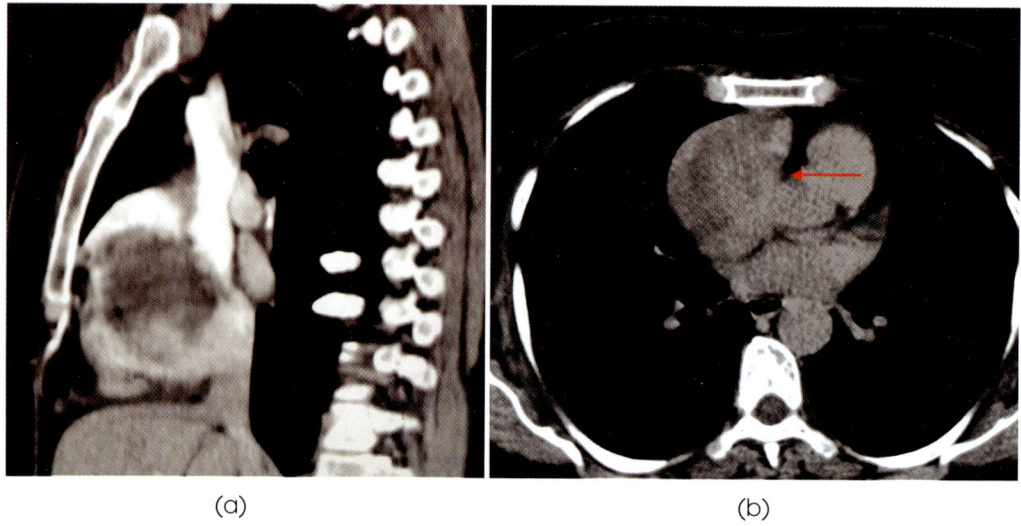

(a)　　　　　　　　　　　　　　　　　　　(b)

Fig. 7.21a and b: CT sagittal reformation showing mass and delayed enhancement image showing persisting enhancement in the mass (arrow)

PROVISIONAL DIAGNOSIS

Benign tumour? Cardiac myxoma.

Surgical Intervention

Excision of the mass was done. Histopathogical examination proved it to be cardiac myxoma.

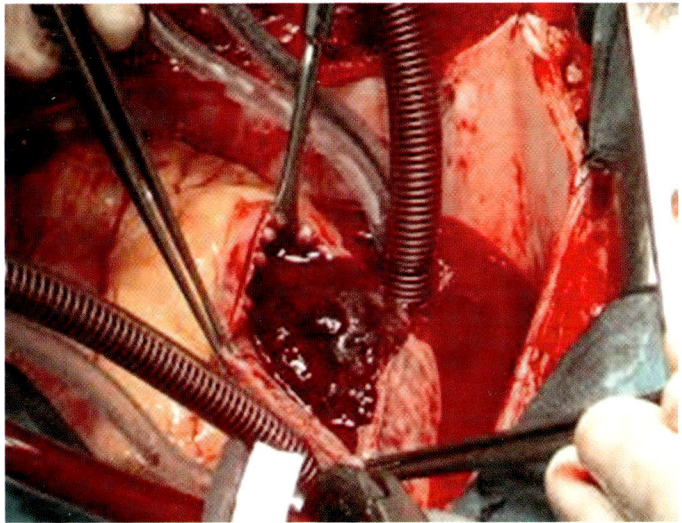

Fig. 7.22: Intraoperative finding

Cardiac Myxoma

- Cardiac myxomas are the most common primary cardiac tumour in adults but are relatively infrequent in childhood.
- There is a broad range in the age of presentation (11–82 years), with most patients presenting in adulthood (mean ~50 years of age).

Clinical presentation

- Approximately 20% of patients are asymptomatic, with myxomas being found incidentally on imaging of the heart. Clinical features are variable and depend on the location and other associated pathology, e.g. valve destruction. However, a triad of symptoms is recognized.
- Valvular obstruction
  - *Left sided:* Dyspnoea, orthopnoea, pulmonary oedema
  - *Right sided:* Symptoms of right heart failure
- Embolic event
  - Distribution will depend on the location of the tumour
  - Most are left sided, and, therefore, most are systemic (brain or extremities)
- Constitutional symptoms
  - Weight loss, fatigue, weakness
  - May resemble infective aetiology (fever, arthralgia, lethargy)
- In ~20% of patients, arrhythmias may be identified

Morphological Appearance

- Location
- Atria (95%)
- Left (75%)
- Right (20%)
- Ventricles (5%)

CT Scan Findings

- They are usually heterogeneously low attenuating (approximately two-thirds of cases).
- Due to repeated episodes of hemorrhage, dystrophic calcification is common.
- Heterogeneous, typically spherical or ovoid masses which may be sessile or pedunculated.

MRI Findings

- T1: Tend to be low to intermediate signal, but areas of hemorrhage may be high
- T2: Can be variable due to heterogeneity in tumour components; e.g. calcific components > low signal; myxomatous components > high signal
- GE (gradient echo): May show blooming of calcific components
- T1 C+ (Gd): Shows enhancement (important discriminator from a thrombus).

Case 7

Patient was k/c of DM/CAD (coronary artery disease) on medication since many years. Patient status is post CABG in 2006. Patient was diagnosed case of pulmonary tuberculosis. He was taking ATT. Now patient got admitted with complaints of altered mental status. MRI brain was done showed multiple infarction. Echocardiography was done showed mass in right pulmonary vein. In view of patient history of pulmonary tuberculosis the differential diagnosis was granuloma or a clot cardiac MRI was done for viability and pulmonary vein lesion evaluation. HRCT chest was also done to see the lung status.

Cardiac MRI and HRCT Chest Findings

- HRCT chest showed diffuse ground glass opacities in both lung fields intermixed with mosaic attenuation, centrilobular densities and patchy consolidation (Fig. 7.23).
- A well-defined hypointense lesion on MRI is seen in right inferior pulmonary vein measuring approximately 2 cm in size. It was protruding in the lumen of left atrium. Rest of all the pulmonary veins were normal. The lesion was not enhancing on post-contrast study (Figs 7.24 and 7.25).

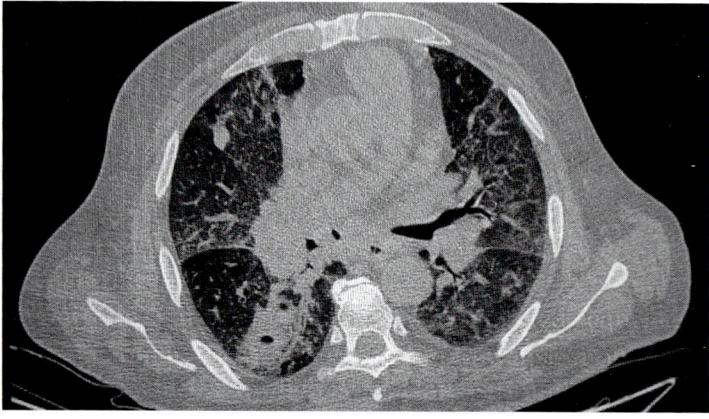

Fig. 7.23: HRCT chest

- Mild hypokinesia of mid-basal-apical inferior wall
- Cardiac MRI contrast (late gadolinium enhancement at interval of 10–15 mins) images showed subendocardial enhancement of mid-basal-apical inferior wall consistent with 25–50% scarring (Fig. 2.26).

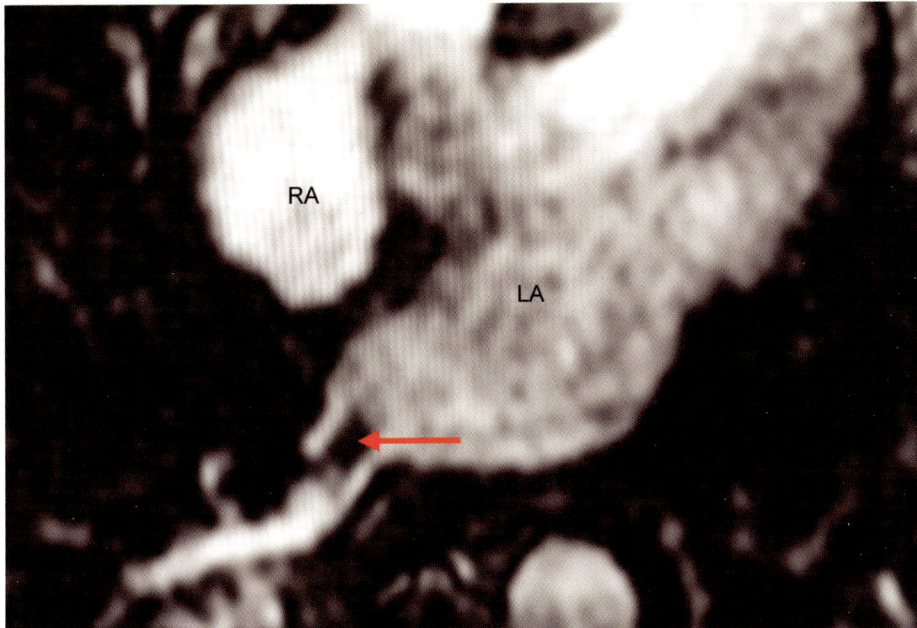

Fig. 7.24: Cardiac MRI showing hypointense lesion in right inferior pulmonary vein (arrow)

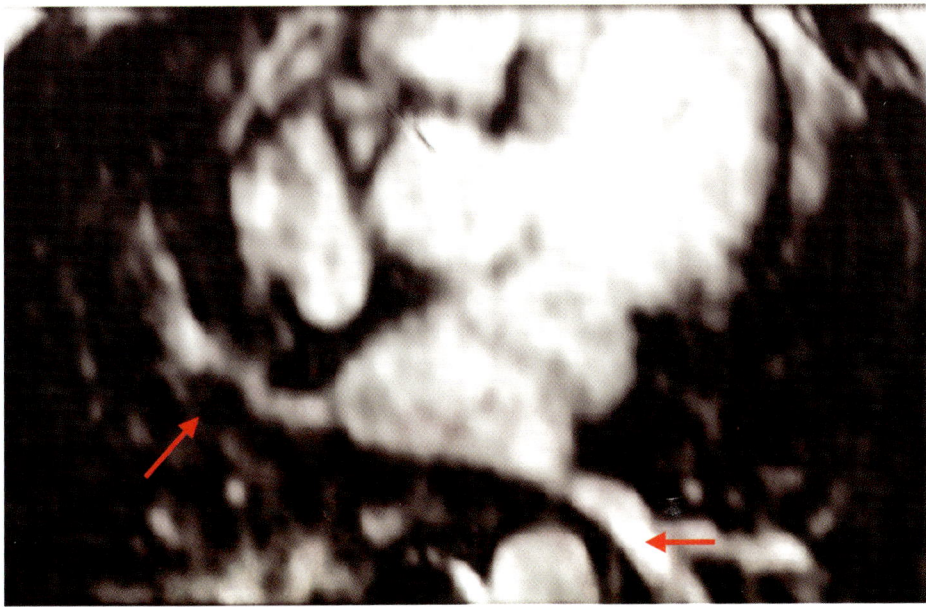

Fig. 7.25: Cardaic MRI showing normal rest pulmanary veins (arrows)

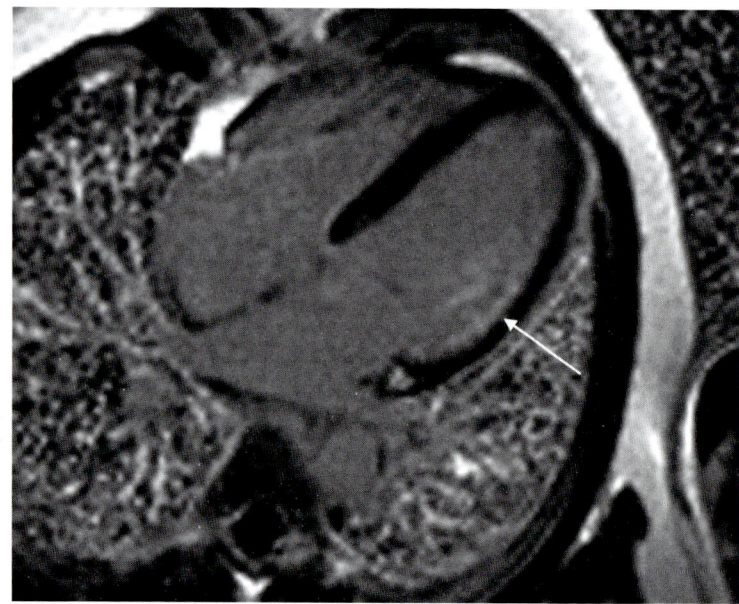

Fig. 7.26: Cardiac MRI short axis image showing subendocardial enhancement (arrow)

Radiological Diagnosis

Clot in right inferior pulmonary vein.

Medical Intervention

Patient was kept on thrombolytic agents and follow-up echocardiography showed resolution of inferior pulmonary vein clot.

Discussion

CMR and cardiac CT currently represent very important imaging modalities used for the comprehensive evaluation of the PV in various clinical settings. Cardiac MR is considered to be the gold standard for volumetric assessment of left atrial size. Late gadolinium enhancement CMR offers unique information about the presence of LA scaring that may be used for optimizing the treatment of patients suffering from atrial arrhythmias. Cardiac magnetic resonance plays a very important role in characterizing PV morphological variants and pathological intra-atrial masses.

Suggested Reading

Pazos-López P., Pozo E., Siqueira M. E., et al. Value of CMR for the differential diagnosis of cardiac masses. JACC Cardiovasc Imaging. 2014; 7(9):896–905. doi: 10.1016/j. jcmg. 2014. 05.

Case 8

Patient is known case of HCC post TACE with extension of tumour thrombus through IVC into right atrium as seen on previous CT scan and echocardiography. Cardiac MRI was done for further evaluation.

Cardiac MRI Findings

- Hypointense mass lesion occupying suprahepatic IVC and cranially extending in right atrium, occupying two-thirds of RV cavity (Figs 7.27a, b and c).
- Post-contrast study shows mild enhancement of the mass (Figs 7.28a and b)
- CINE images show no obvious wall motion abnormality of LV/RV wall.
- No delayed enhancement of LV myocardium suggesting viable myocardium

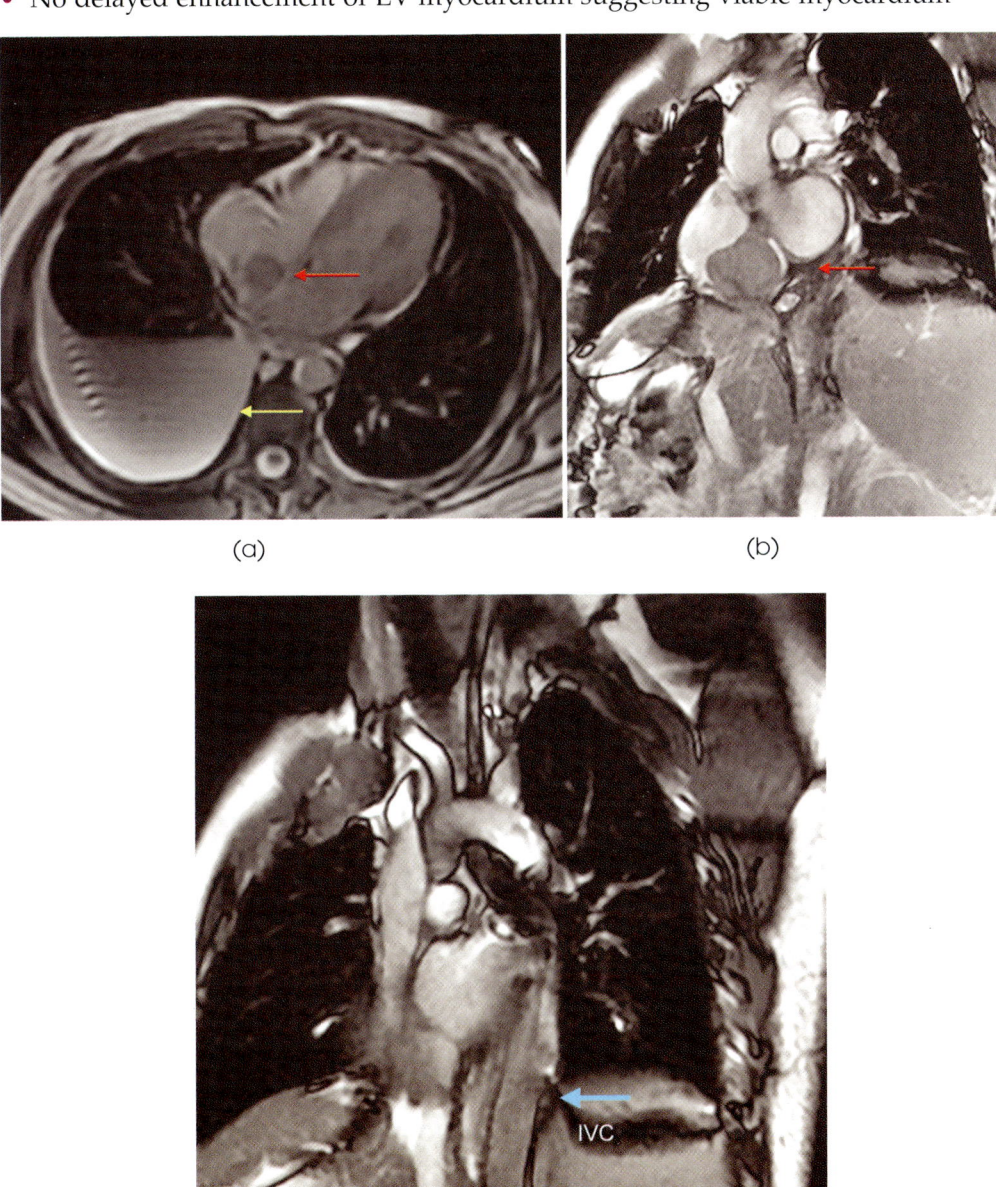

(a)　　　　　　　　　　　　　　(b)

(c)

Fig. 7.27a, b and c: Cardiac MRI axial and coronal fiesta images showing tumour in right atrium (red arrows), tumour extension through IVC (blue arrow) and right pleural effusion (yellow arrow).

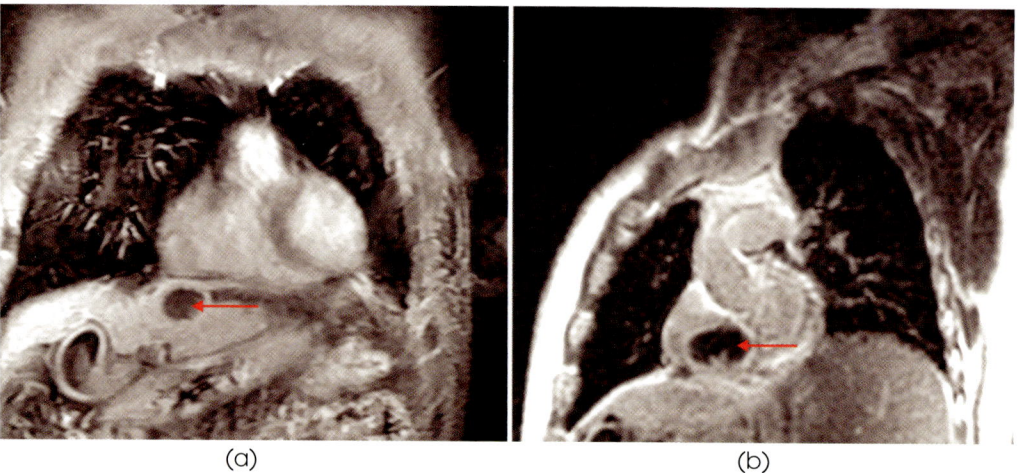

(a)   (b)

Fig. 7.28a and b: Post-contrast coronal images showing mild peripheral enhancement (arrows)

Radiological Diagnosis

Tumour thrombus in right atrium and suprahepatic IVC.

Surgical Intervention

Patient underwent for hepatic transplant and removal of tumour from IVC and right atrium.

Discussion

- The incidence of portal vein tumour thrombus in HCC has been reported to range from 6.5% to 44%
- invasion of a tumour thrombus into the inferior vena cava and right atrium is an infrequent occurrence, seen in about 3% to 4% of patients with HCC.

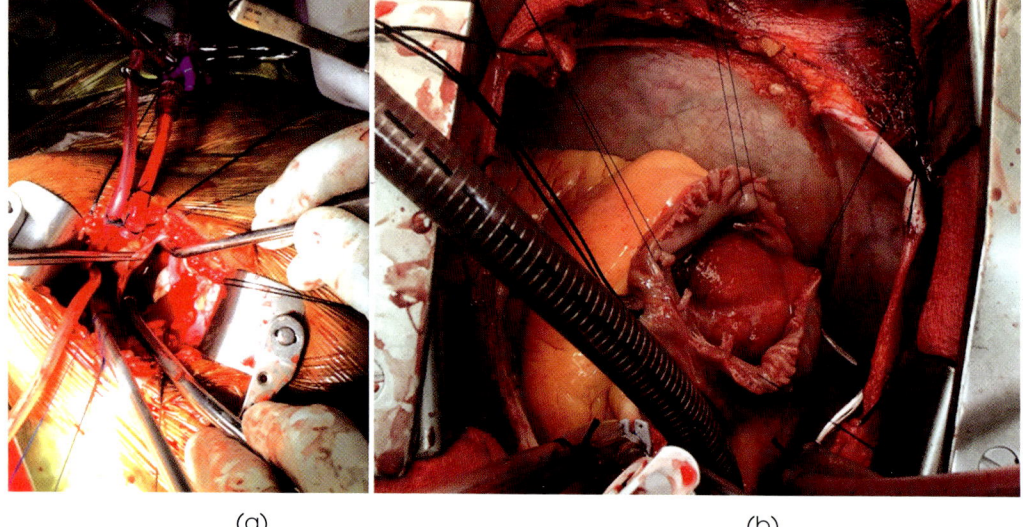

(a)   (b)

Fig. 7.29a and b: Peroperative figures showing mass

- Cavoatrial tumour thrombus occurs as an extension of the tumour from one or more of the hepatic veins. Patients with HCC, who present with a cavoatrial tumour thrombus have a dismal prognosis, with the survival in this setting being less than 3 months without treatment.
- Complications specific to the presence of the cavo-atrial tumour thrombus include Budd-Chiari syndrome, pulmonary embolism, heart failure and sudden cardiac arrest.

Case 9

Patient is known case of renal cell carcinoma (RCC).

Contrast CT Thorax Findings

Biopsy proven RCC with extension of mass in IVC and intra-atrial tumour thrombus formation (Figs 7.30a, b, c and d, and 7.31).

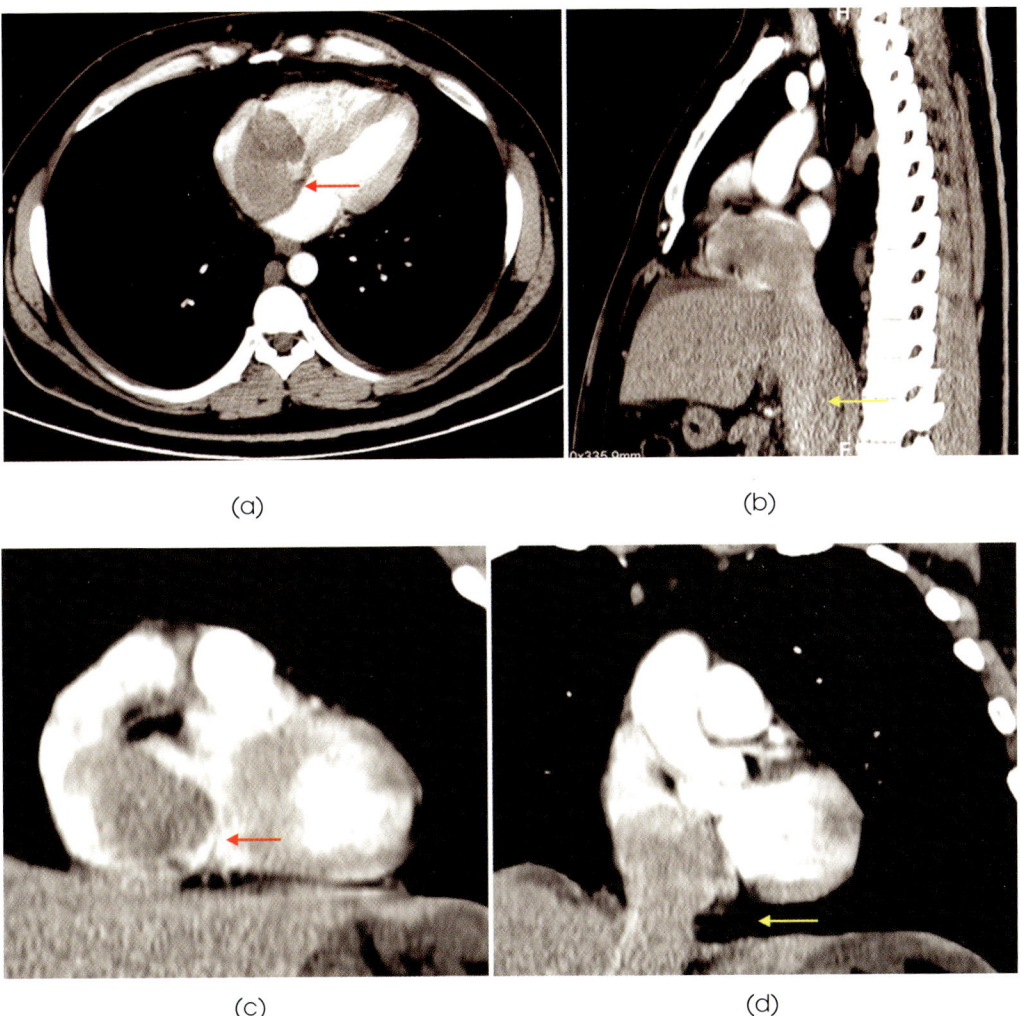

(a)　　　　　　　　　　　　　　　　(b)

(c)　　　　　　　　　　　　　　　　(d)

Fig. 7.30a, b, c, and d: Contrast CT image showing hypodense mass lesion in right atrium (red arrow) and visualised IVC (yellow arrow)

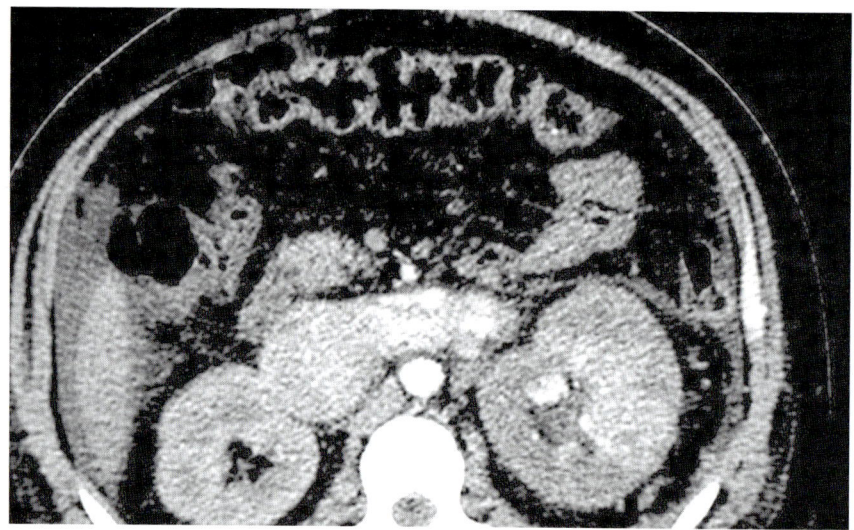

Fig. 7.31: Contrast CT image showing left renal mass and tumour thrombus in IVC

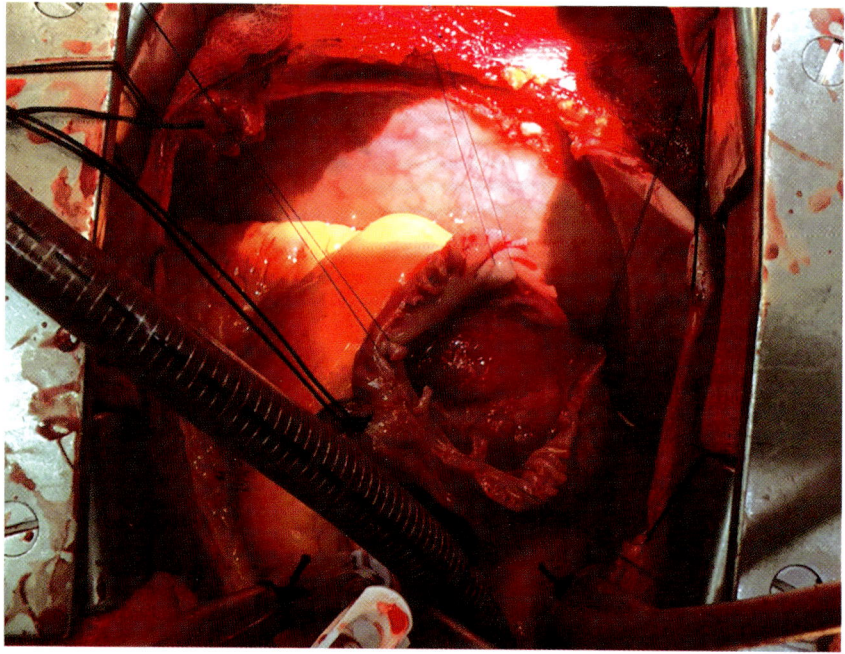

Fig. 7.32: Peroperative figure showing tumour

Radiological Diagnosis

Biopsy proven renal cell carcinoma left kidney with now tumour extension through IVC into right atrium.

Surgical Intervention

Patient underwent surgical removal of left kidney and simultaneously removal of right atrial mass and IVC tumour thrombus (Fig. 7.32) peroperative.

Discussion

Renal cell carcinoma extends into the IVC in 4 to 10% of all the patients. IVC involvement in RCC usually occurs by intraluminal extension of the tumour. It can if left untreated, lead to poor survival and quality of life. The 5-year survival rates of 47 to 68% have been reported following complete surgical excision of localized RCC in this setting. Factors that finally influence the outcome in these patients include clinical staging, completeness of resection, and biological characteristics of the primary tumour.

Suggested Reading

Ciancio G, Soloway M. Resection of the abdominal IVC for complicated renal cell carcinoma with tumour thrombus. BJU Int. 2005; 96: 815–8.

Left Ventricular Aneurysm *versus* Pseudoaneurysm

Patient is known case of DM, HT and triple vessel disease with left ventricle failure symptoms.

On echocardiography a large aneurysm was seen arising from LV posterior-basal segment (Fig. 8.1). As patient had high creatinine value so noncontrast MRI was done.

Cardiac MRI Findings

1. Dilated left ventricular large aneurysm arising from posterio basal wall of left ventricle measuring 7 × 5 cm size with another bulge in the aneurysm wall posteriorly and thinning of wall in this region with possibly dehiscence in endocardium and surrounded by myocardium (Figs 8.2a and b).
2. Reduced left ventricular systolic function
3. Moderate to severe MR

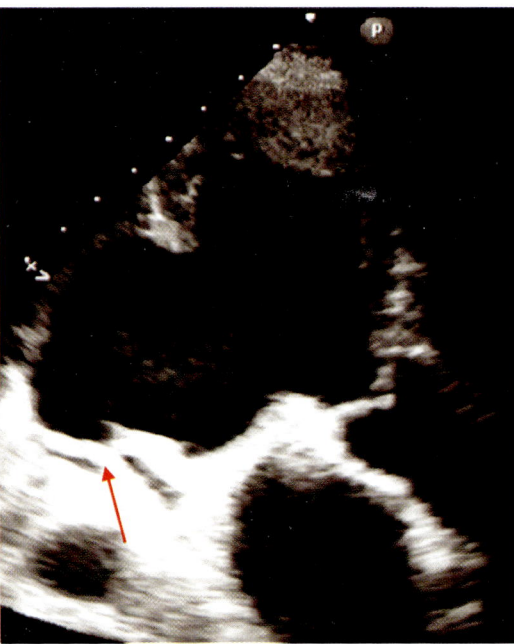

Fig. 8.1: Echocardiography showing large aneurysm arising from posterior-basal wall

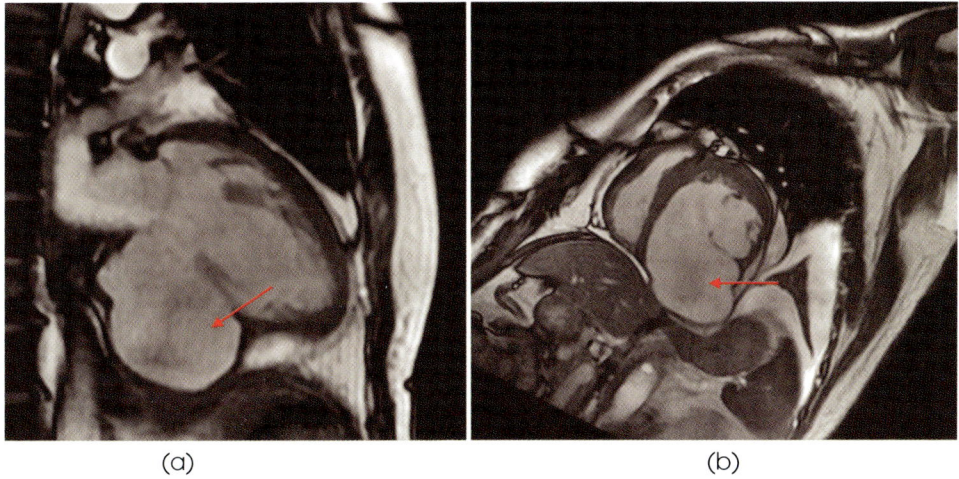

(a)   (b)

Fig. 8.2a and b: Cardiac MRI showing large aneurysm with wide mouth connection with left ventricle

Radiological Diagnosis
LV true aneurysm

X-ray/CT Scan/MRI Features of True and False Aneurysm
True Aneurysm
- Localized out pouching of ventricular cavity
- Wide-mouthed in connection with the LV
- Frequently not visible on chest X-ray but may produce localized bulge of left heart border
- Paradoxical expansion during systole
- May have rim of calcium in fibrotic wall.

False Aneurysm
- False aneurysm occurs when left ventricle ruptures into pericardial sac.
- Pericardial adhesions contain rupture
- Diameter of mouth is smaller.

Causes
- Myocardial infarction (MI)
- Trauma.

Surgical Intervention
- CABG with mitral valve replacement (MVR) and aneurysm repair, then patient kept on IABP and high inotropic support.
- Patient was shifted back to CTVS ICU on ventilator with ET tube *in situ* and all invasive lines *in situ* on IABP with moderate to high inotropic and vasopressor
- Support of EPI/NOREPI/DOBU/milrinone/vasopression was given.

Case 2

Patient is status post-myocardial infarction (MI) VSD repair now complaints of aneurysmal dilatation of left ventricle as seen on echocardiography now for further evaluation underwent cardiac MRI.

Cardiac MRI Findings

- Dilated cardiomyopathy
- Left ventricular pseudoaneurysm arising from inferio-basal segment having small mouthed connection with left ventricle (Fig. 8.3).
- Enhancement of wall of pseudoaneurysm which is formed by fibrosed pericardial layer (Fig. 8.4).

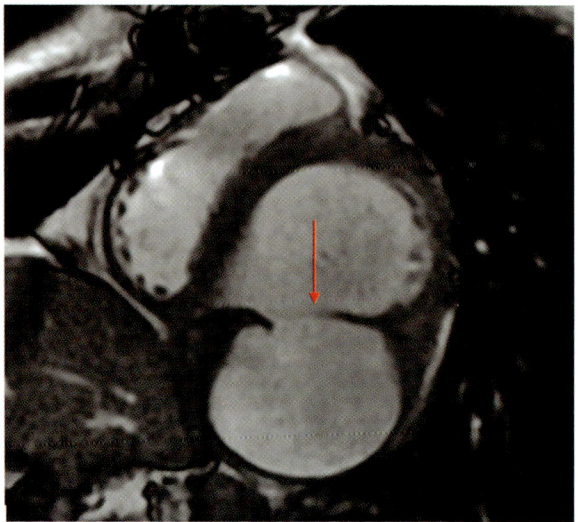

Fig. 8.3: Cardiac MRI 2-chamber showing left ventricular pseudoaneurysm arising from inferio-basal segment having small mouthed connection with left ventricle (arrow)

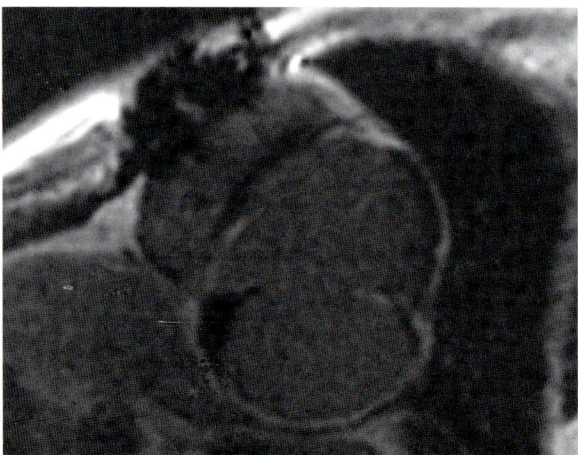

Fig. 8.4: Cardiac MRI post-contrast image showing enhancement of the wall of the pseudoaneurysm

Radiological Diagnosis

Left Ventricular Pseudoaneurysm

False or pseudoaneurysms: False aneurysm occurs when left ventricle ruptures into pericardial sac. It has smaller diameter of mouth. Pericardial adhesions contain rupture usually has no calcium rim.

Left Ventricular Assist Device

Case 1

Patient is 38-year-old male known case of dilated cardiomyopathy with pacemaker for a few years now increasing breathlessness. Patient was with multiorgan failure, dearranged KFT and LFT. Echocardiography showed dilated cardiomyopathy with very reduced LV function (EF—20%).

CT and CXR chest were done which showed bilateral extensive pulmonary oedema (Figs 9.1 and 9.2) respectively.

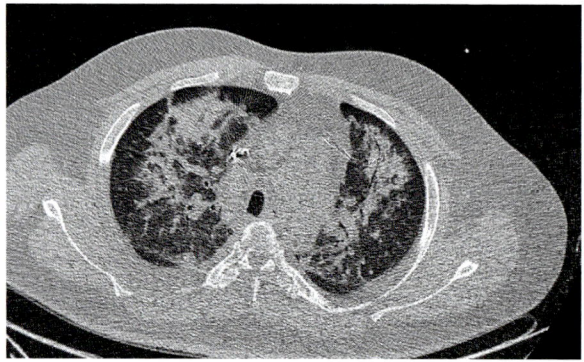

Fig. 9.1: CT showing bilateral pulmonary oedema

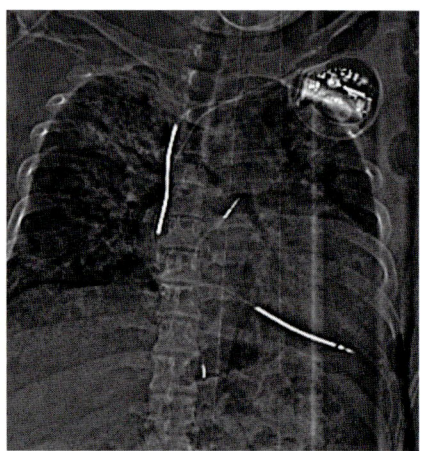

Fig. 9.2: CXR prior to surgery showing bilateral pulmonary oedema

Left ventricular assist device (LVAD) was inserted and patient's clinical condition was improved. Then at the time of discharge CT aortogram was advised as per protocol.

The Findings Were

- Blood was seen taken away via an apical inflow cannula.
- The blood moves through the device to the outflow graft that typically returns blood via an anastomosis to the mid-ascending aorta. (Fig. 9.3)
- Normal outflow graft—position is defined as insertion into ascending aorta prior to innominate artery origin. No kinks were seen (Fig. 9.4).
- Normal inflow cannula position—IV apical location with cannula tip not directly abutting LV wall, i.e. without obstruction to inflow.
- Postoperative CXR was done which shaved no evidence of pulmonary oedema with normally placed LVAD tip.

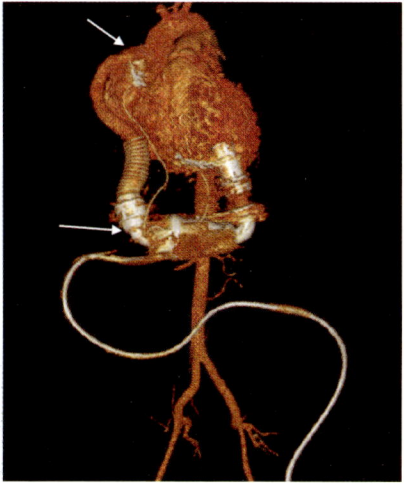

Fig. 9.3: Normal inflow cannula position

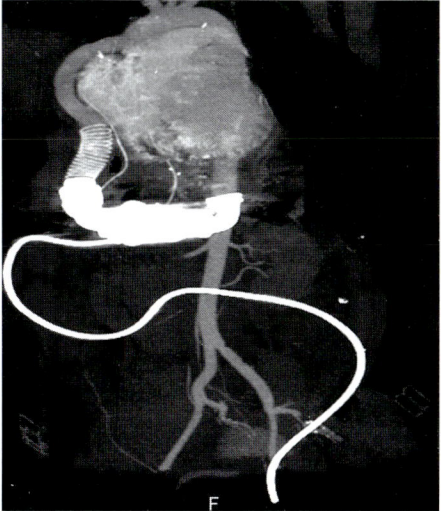

Fig. 9.4: Normal outflow graft

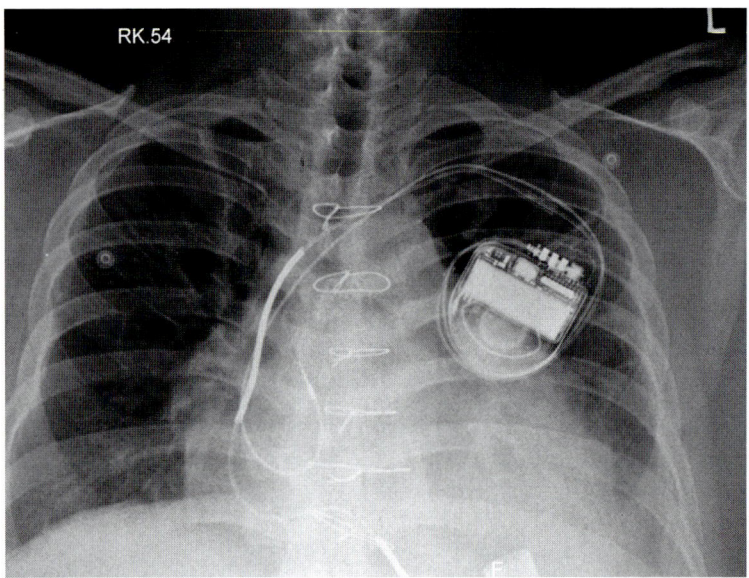

Fig. 9.5: Post LVAD normal CXR showing pacemaker and LVAD inflow cannula tip

Indications of CT Angiography in LVAD

1. Low cardiac output symptoms.
2. Assessment of cannula position.
3. Low flow reading on LVAD monitor.
4. Surgical planning

Conclusion

In patients with severe heart failure, LVADs have become a more widely used treatment option as a bridge to heart transplantation, as destination therapy, and as a bridge to myocardial recovery. It is important to have an understanding of the components of the devices and their normal imaging features. Such knowledge can also prepare radiologists to recognize the imaging features of complications arising from these devices.

Suggested Reading

1. DeBakey ME. Development of mechanical heart devices. Ann Thorac Surg 2005; 79(6): S2228–S2231. Cross Ref, Medline.
2. Clegg AJ, Scott DA, Love man E, et al. The clinical and cost-effectiveness of left ventricular assist devices for end-stage heart failure: a systematic review and economic evaluation. Health Technol Assess 2005; 9(45): 1–132.

Myocardial Infarction (MI), Dysfunction and Coronary Artery Disease

Case 1

Patient is a 45-year-old male having history of coronary artery disease with anterior wall MI. On echocardiography ejection fraction was low. There was mild hypokinesia of anterior wall of left ventricle. An echogenic layer was seen in left ventricular apical region.

Patient was advised for MRI for further evaluation.

Cardiac MRI Findings

1. Nonenhancing clot was lying in ventricular apex which was hypointense on all sequence (Figs 10.1a and b).
2. There was transmural enhancement consistent with non-viable myocardium involving ventricular apex and adjoining myocardium (Fig. 10.2).
3. There was subendocardial enhancement in mid-basal septum suggesting 25% wall thickness nonviable myocardium. Rest of the myocardium was viable.
4. Left ventricular qualitative analysis showed reduced systolic function.
   – Regional wall motion abnormality analysis showed akinetic LV apical region and hypokinesia of mid-basal septum.
   – Rest of the myocardium showed normal motion.

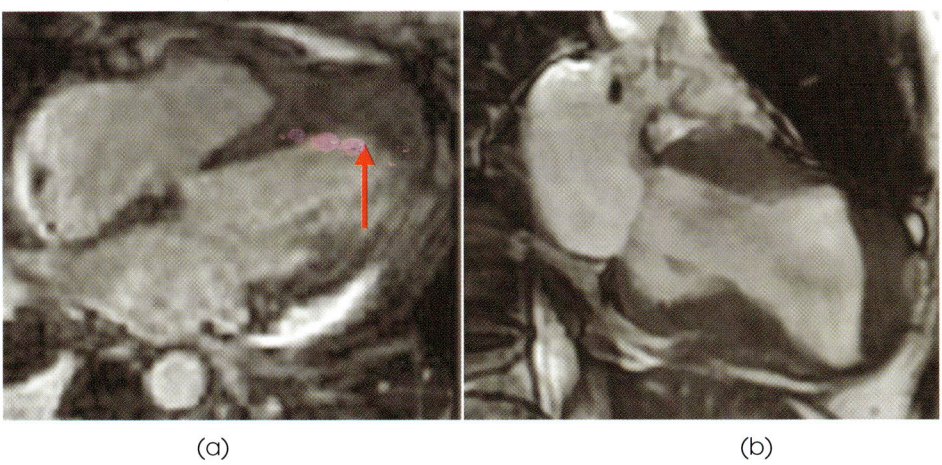

(a)   (b)

Figs 10.1a and b: Cardiac MRI showing layer of clot in left ventricular apex (arrow)

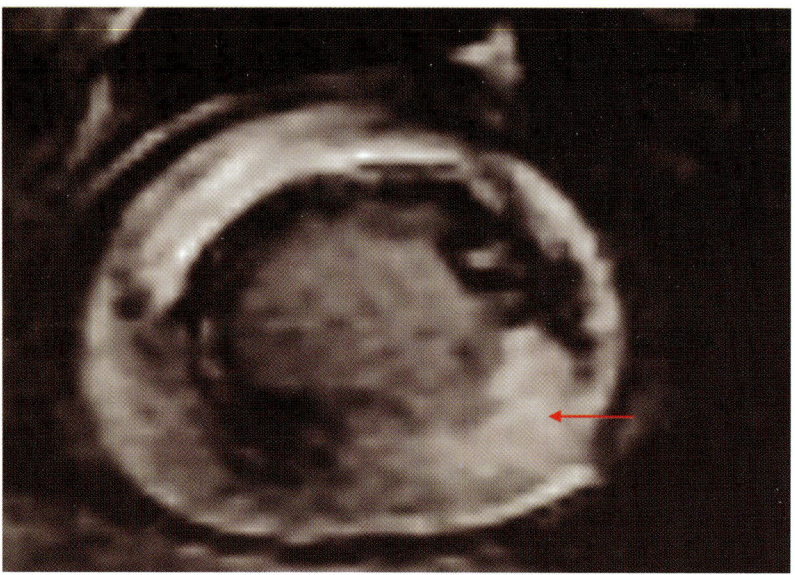

Fig. 10.2: Contrast cardiac MRI showing transmural enhancement consistent with non-viable myocardium involving ventricular apex and adjoining myocardium (arrow)

Surgical Intervention

LV clot removal and stenting of left anterior descending artery was done.

Conclusion

Cardiovascular MRI provides a unique tool to assess multiple interrelated clinical markers of viability in a single test. Its overall accuracy appears to be equivalent, and in several reports, superior to the currently available techniques, including PET imaging. Considering the greater spatial resolution compared with PET and the wealth of correlative pathological data, DE-MRI may well represent the new gold standard in the detection of irreversibly damaged myocardium. However, the clinical data to date consist of relatively small numbers of patients, and setting a convincing new standard will require larger and more definitive clinical trials. Nonetheless, it is apparent that the full potential of CMR has only just begun to emerge, and its impact on the management of LV dysfunction will continue to increase.

Suggested Reading

1. Rahimtoola SH. A perspective on the three large multicenter randomized clinical trials of coronary bypass surgery for chronic stable angina. Circulation. 1985; 72 (6 pt 2): V-123-V-135. Medline.
2. Braunwald E, Kloner RA. The stunned myocardium: prolonged, postischemic ventricular dysfunction. Circulation. 1982; 66: 1146–1149.

Case 2

A 40-year-old male with complaints of breathlessness on exertion. Normal LV function on echocardiography with LV hypertrophy. Cardiac MRI was done to rule out myocardial dysfunction/dysarray.

MRI Findings

1. Asymmetric ventricular wall thickening. Normal atrial chambers.
2. Global mild hypokinesia of LV wall.
3. LGE shows patchy enhancement in mid-septal, inferio-basal and lateral wall.
4. Normal LV systolic function. LVEF—76%, LVEDV—108.8 ml, LVESV—25.5 ml CO—5.7 l/min
5. No LVOT obstruction with normal aortic flow.
6. MRI tagging suggesting myocardial dysarray with reduced myocardial strain in basal region.

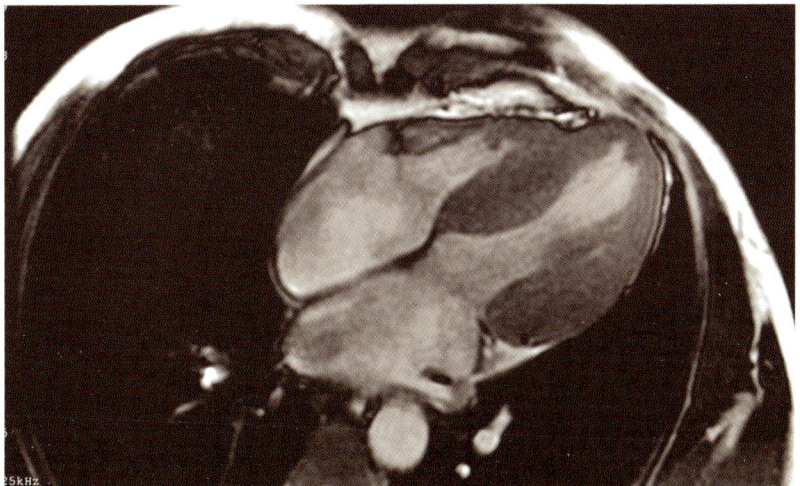

Fig. 10.3: Cardiac MRI 4-chamber showing concentric left ventricular hypertrophy

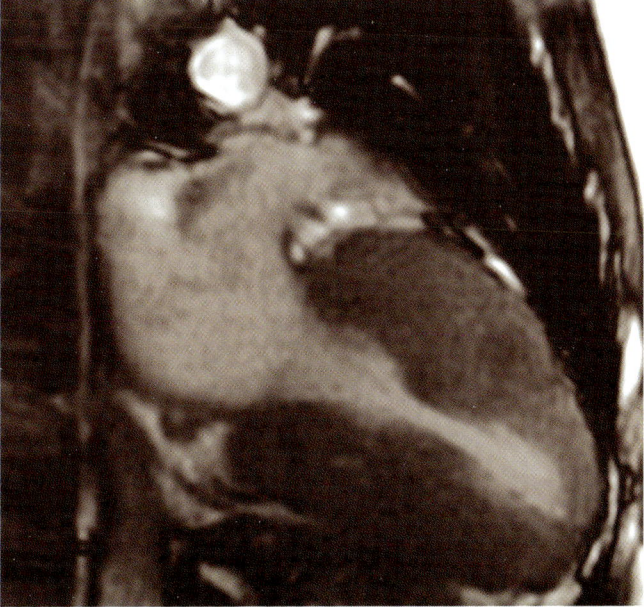

Fig. 10.4: Cardiac MRI showing LV hypertrophy

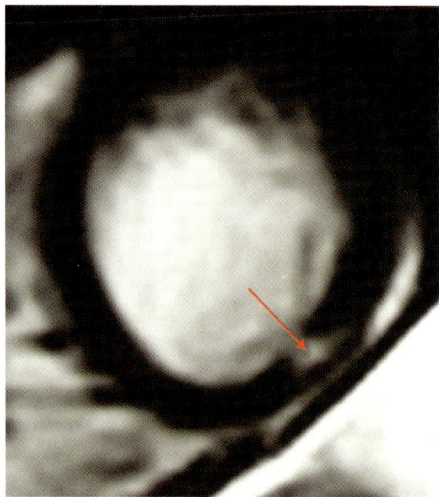

Fig. 10.5: Contrast MRI showing patchy enhancement in LV myocardium (arrow)

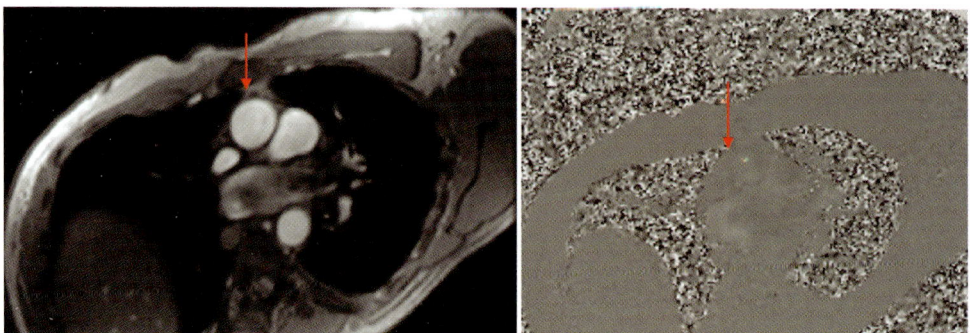

Fig. 10.6: Venc study for aortic flow analysis (arrow)

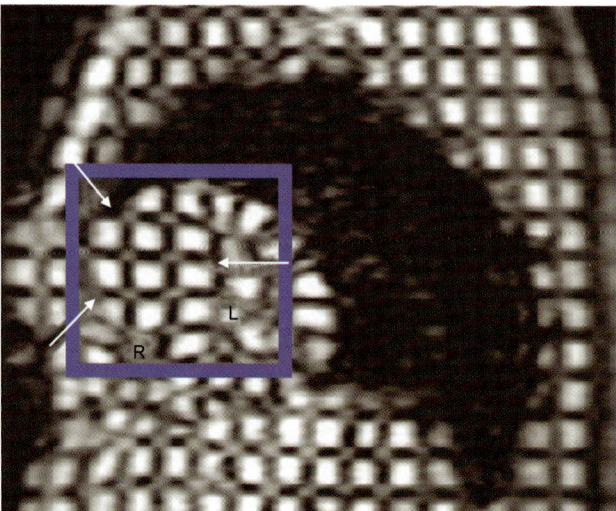

Fig. 10.7: Cardiac MRI tagging showing region of akinesis identified by undeformed grid lines and surrounded by deformed grid lines (arrows)

Radiological Diagnosis

Asymmetric ventricular wall thickening with normal atrial chambers, global mild hypokinesia of LV wall, patchy enhancement in mid-septal, inferio-basal and lateral wall, normal LV systolic function. LVEF—76%, lVEDV—108.8 ml, lVESV—25.5 ml, CO—5.7 l/min, 5. No LVOT obstruction with normal aortic flow and MRI tagging suggesting myocardial dysarray with reduced myocardial strain in basal region.

Discussion

Myocardial tagging is an advanced CMRI technique for measuring regional heart function. This technique provides virtual markers for quantifying myocardial deformation (e.g. strain, strain rate and torsion).

Conclusion

CMR tagging is powerful non-invasive diagnostic tool for quantifying regional and diastolic function due to its multiplanar capability, and in early detection and management of wide range of myocardial disease.

Suggested Reading

1. Amundsen BH, Helle-Valle T, Edvardsen T, Torp H, Crosby J, Lyseggen E, Støylen A, Ihlen H, Lima JAC, Smiseth OA, Slørdahl SA: Noninvasive Myocardial Strain Measurement by Speckle Tracking Echocardiography: Validation Against Sonomicrometry and Tagged Magnetic Resonance Imaging. Journal of the American College of Cardiology. 2006, 47 (4): 789–793.
2. Edvardsen T, Gerber BL, Garot J, Bluemke DA, Lima JA, Smiseth OA: Quantitative assessment of intrinsic regional myocardial deformation by Doppler strain rate echocardiography in humans: validation against three-dimensional tagged magnetic resonance imaging. Circulation. 2002, 106 (1): 50–56.

Case 3

Patient is 63-year-old male status post-tripple vessel disease (TVD).

Cardiac MRI was done for myocardial viability, motion analysis and left ventricular function.

Cardiac MRI Findings

- Dilated LA and LV (Fig. 10.8)
- Nonviable left ventricular apex 50–75% thickness
- Sub-endocardial enhancement of anterioseptal and inferioseptal wall consistent with 25% wall thickness nonviable myocardium (Fig. 10.9)
- Reduced LV systolic function EF—25%
- Akinetic apex and hypokinesia of rest of LV wall (Fig. 10.10).

Radiological Diagnosis

Nonviable LV apex 50–75% thickness, subendocardial enhancement of anterioseptal and inferioseptal wall consistent with 25% wall thickness nonviable myocardium and reduced LV systolic function EF—25%.

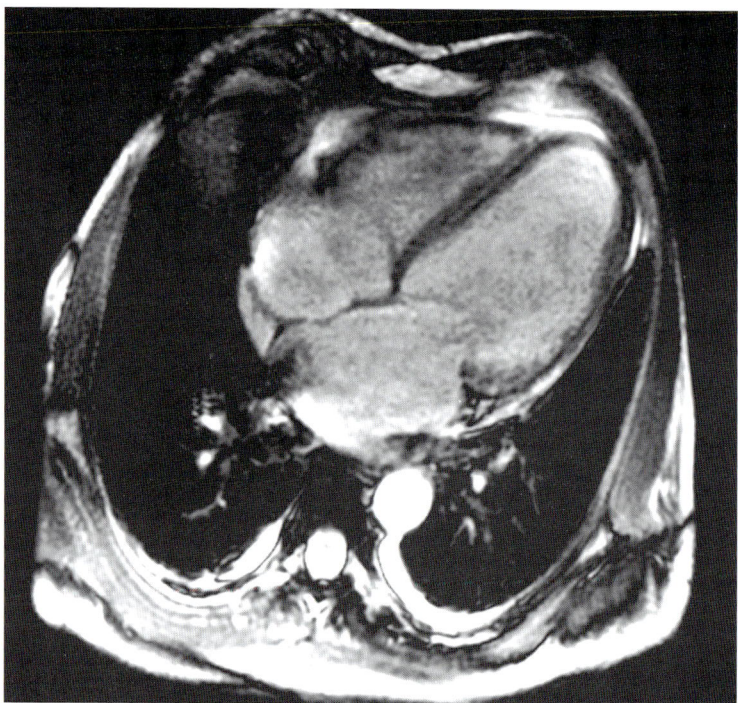

Fig. 10.8: Cardiac MRI showing dilated left ventricle

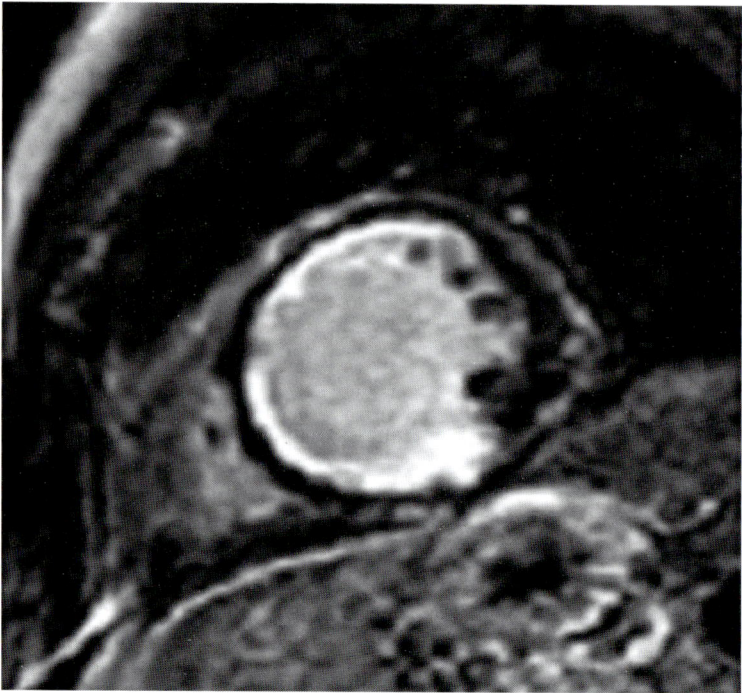

Fig. 10.9: Cardiac MRI showing subendocardial enhancement of anterioseptal and inferioseptal wall consistent with 25% wall thickness nonviable myocardium

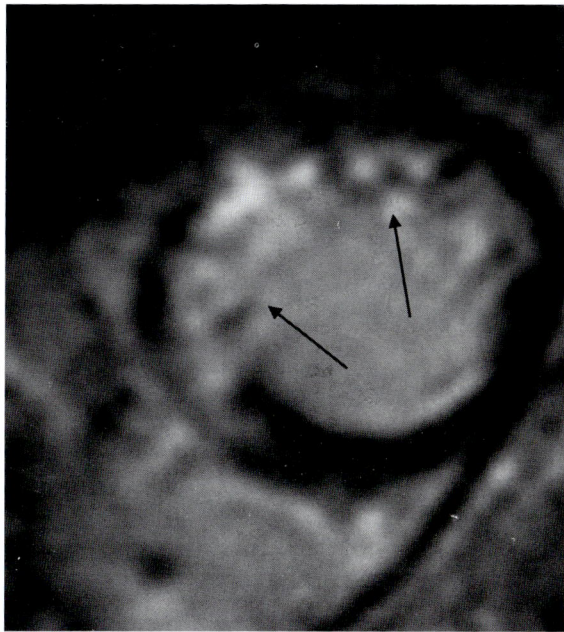

Fig. 10.10: MRI showing nonviable left ventricular apex 50–75% thickness (arrows)

Discussion

Cardiovascular MR (CMR) is a rapidly emerging non-invasive imaging technique, providing high resolution images of the heart. The CMR technique has been named delayed-enhancement (DE-MRI) and demonstrate nonviable tissue as "hyperenhanced" or bright. In severe ischemic heart failure, MRI hyper enhancement as a marker of myocardial scar closely agrees with PET data.

Advantages of Cardiac MRI

- Greater availability of the equipment and trained manpower.
- No ionizing radiation.
- MRI is cheaper than PET.
- No radiopharmaceutical agents.

Suggested Reading

1. Flacke SJ, Fischer SE, Lorenz CH. Measurement of the gadopentetatedimeglumine partition coefficient in human myocardium *in vivo:* normal distribution and elevation in acute and chronic infarction. Radiology. 2001; 218: 703–710.
2. Kim RJ, Fieno DS, Parrish TB, et al. Relationship of MRI delayed contrast enhancement to irreversible injury, infarct age, and contractile function. Circulation. 1999; 100: 1992–2002.

Case 4

Patient status is post-coronary artery bypass grafting (CABG) with LIMA to LAD and SVG to OM now having complaints of chest pain. CT coronary angiography was done for graft analysis.

CT Angiography Findings

- LIMA to LAD—patent left internal mammary artery to left anterior descending artery (Fig. 10.11).
- SVG to OM (Saphenous venous graft to obtuse marginal)—small nipple of contrast in the ascending aorta at site of proximal anastomosis with a few small foci of contrast seen around this site. No evidence of contrast beyond this—thrombosis with complete occlusion of graft (Fig. 10.12).

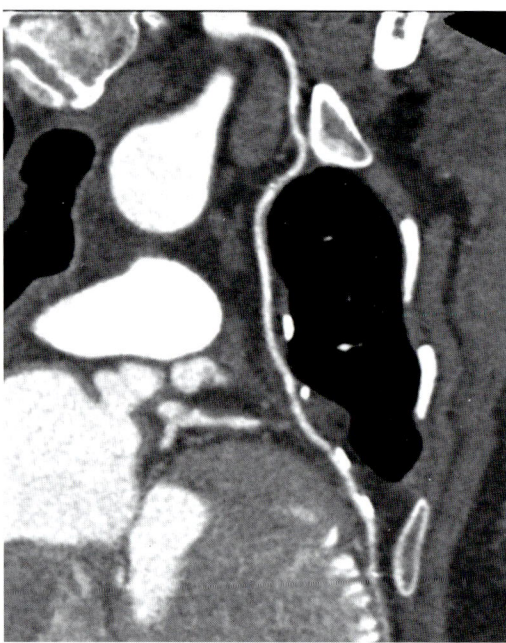

Fig. 10.11: CT angiography showing patent LIMA to LAD

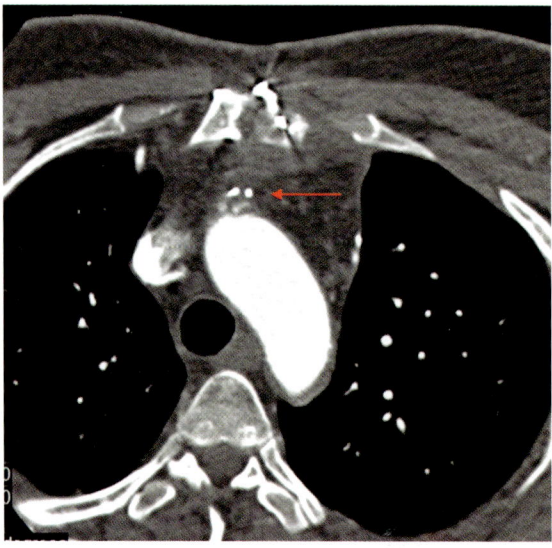

Fig. 10.12: CT angiography showing blocked SVG to OM nipple of contrast is seen near the site

- Native LAD has 90% stenosis proximal to anastomotic site (Fig. 10.13)
- Circumflex (Cx) artery had mild stenosis (Fig. 10.14)
- RCA (right coronary artery) was normal in calibre (Fig. 10.14).
- Volume rendering images (Figs 10.15 and 10.16).

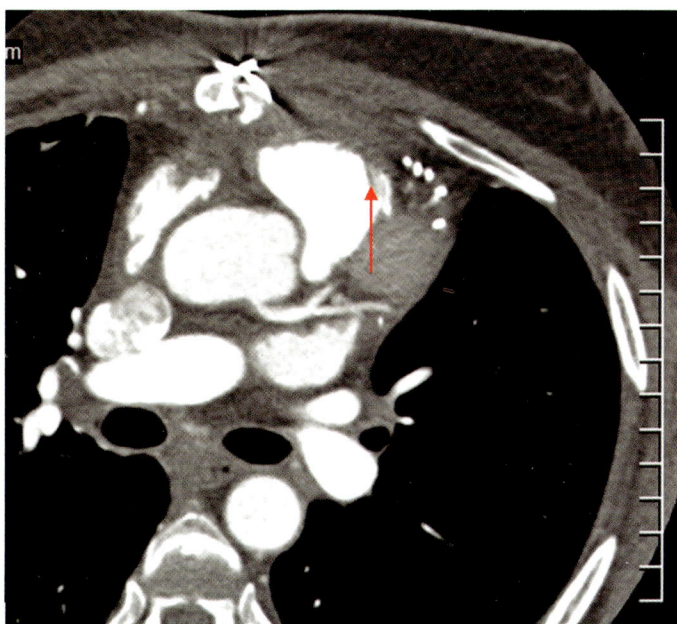

Fig. 10.13: Native LAD severe stenosis proximal to stenosis (arrow)

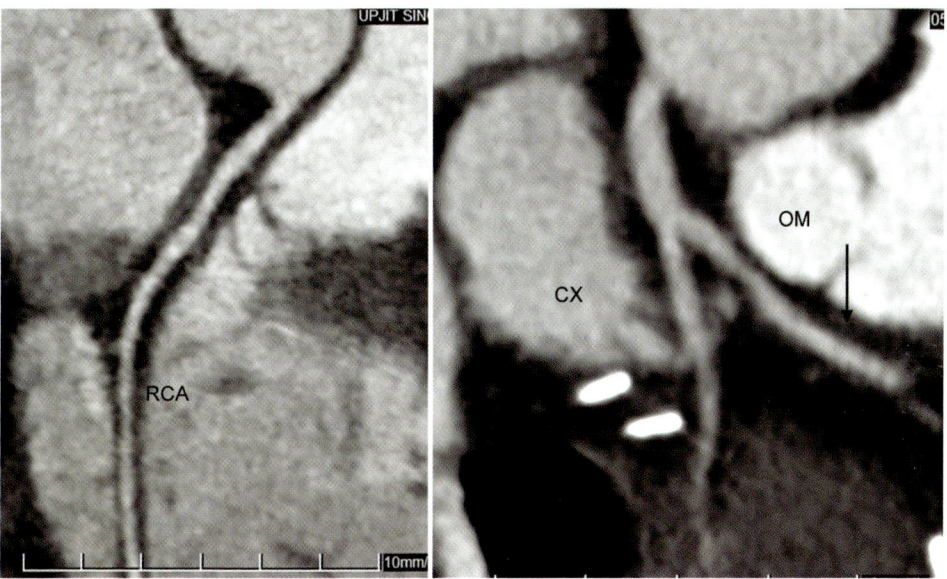

Fig. 10.14: CT angiography showing native RCA, CX and OM, there is complete stenosis of mid-OM (arrow)

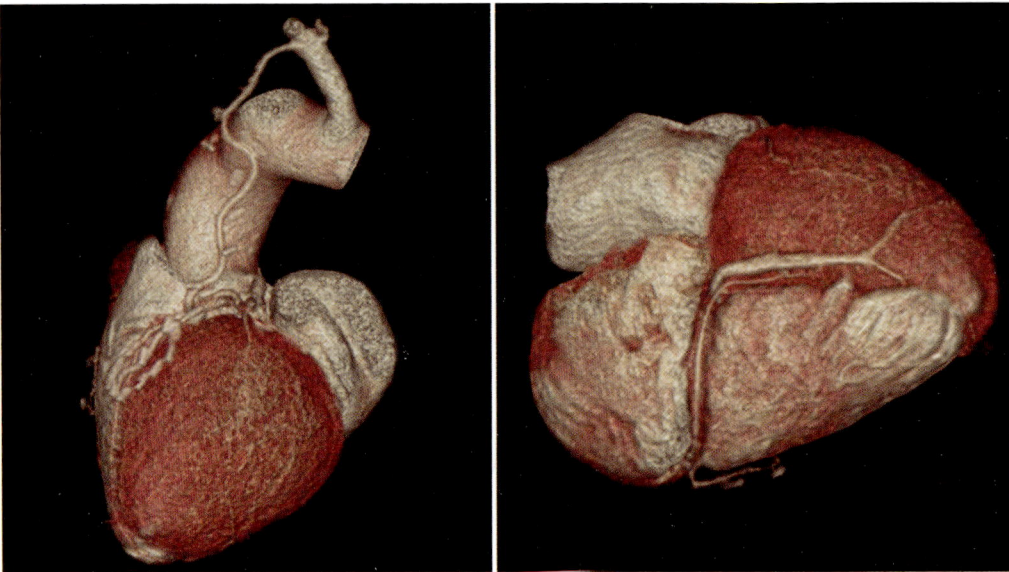

Fig. 10.15: Volume rendering images

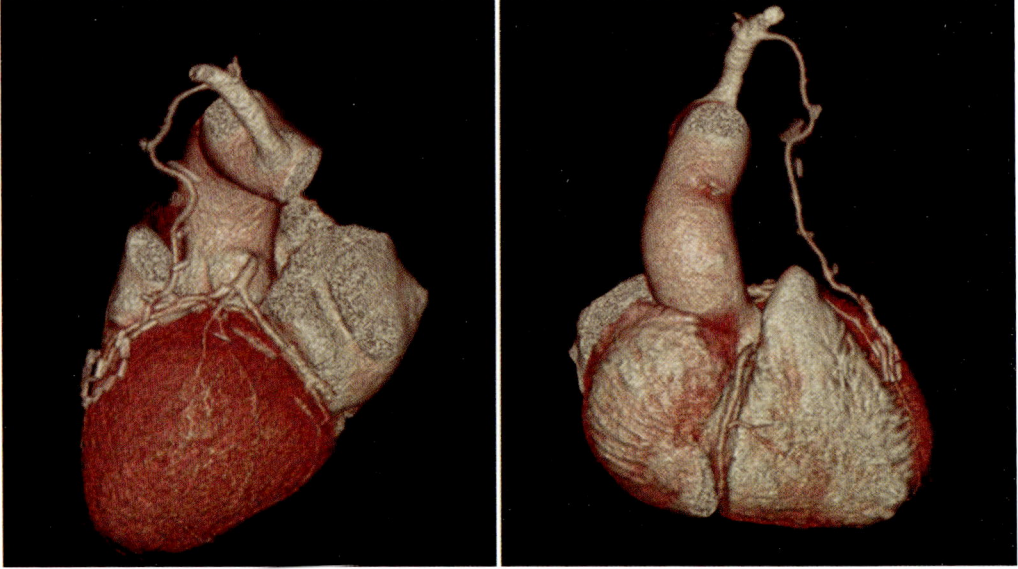

Fig. 10.16: Volume rendering image

Radiological Diagnosis

Patent LIMA to LAD and blocked SVG to OM at ostium due to thrombus formation.

Medical Management

Patient was kept on antithrombotic agents and repeat catheter angiography showed normal ostium and whole course of SVG to OM

Mediastinal and Other Mediastinal/Chest Wall Lesions

Case 1

Patient is 17-year-old girl, a follow-up case of mediastinal mass. Histopathology in her country came out to be granulomatous disease for which she took Koch's treatment first for 3 months, then stopped and then again for one year. Patient approached to cardiologist in hospital and referred for chest MRI and CT chest here.

CT Chest and Cardiac MRI Findings

CT/MRI chest and cardiac MRI showed large diffuse soft tissue density lesion occupying mediastinum obliterating fat planes encasing mediastinal vessels, trachea and bronchi. There was moderate pericardial effusion with air fluid level (Figs 11.1, 11.2 and 11.3). Cardiac MRI showed large right atrium, hypertrophied LV measuring 2.5 cm (end diastolic wall thickness) and heterogeneous enhancement of LV myocardium (Figs 11.4a, b and 11.5).

Radiological Diagnosis

Churg-Strauss syndrome (eosionophillic granulomatosis with polyangitis)

• Churg-Strauss syndrome a rare, small-sized vessel necrotizing vasculitis—also includes coronary arteries.

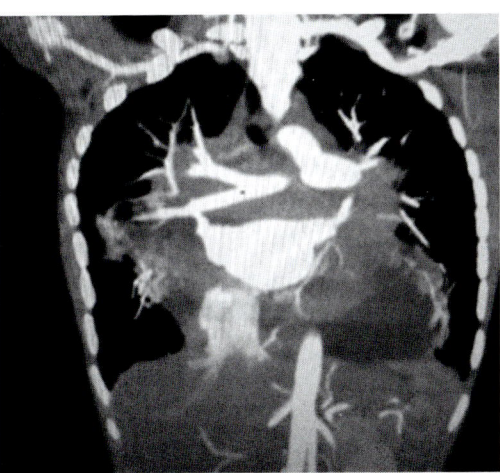

Fig. 11.1: CECT chest showing large diffuse soft tissue density lesion occupying mediastinum obliterating fat planes encasing mediastinal vessels, trachea and bronchi. There is moderate pericardial effusion with air fluid level

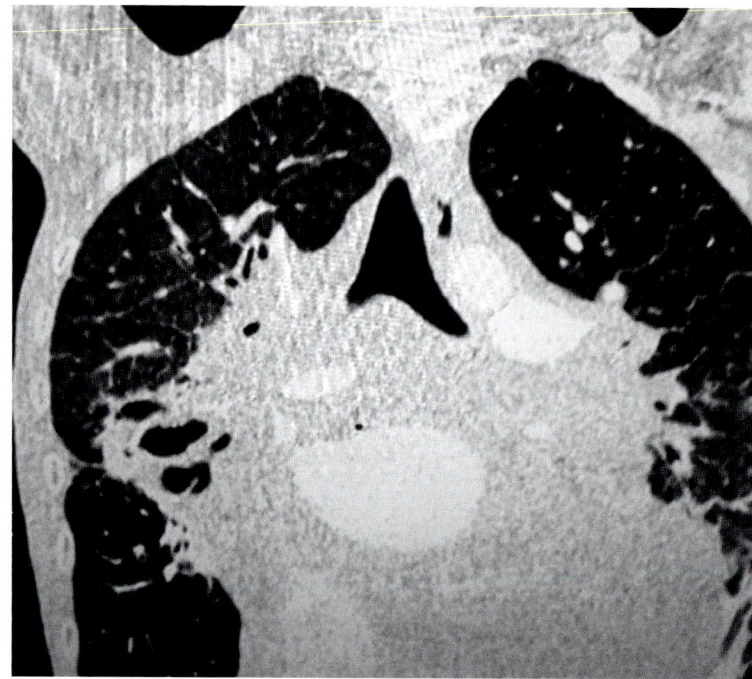

Fig. 11.2: CECT chest lung window showing encasement of bilateral bronchi

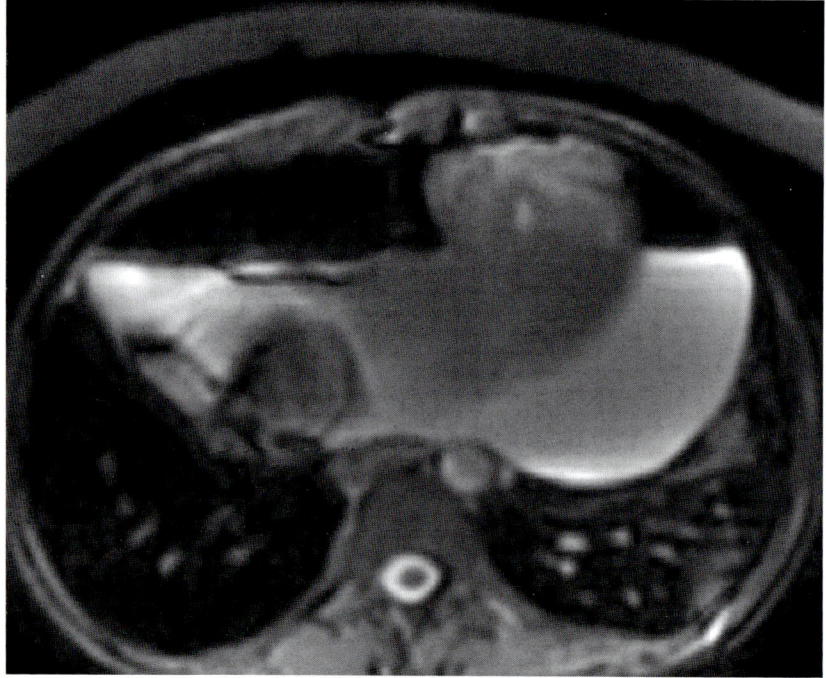

Fig. 11.3: MRI chest showing large diffuse soft tissue intensity lesion occupying mediastinum obliterating fat planes encasing mediastinal vessels, trachea and bronchi. There is moderate pericardial effusion with air

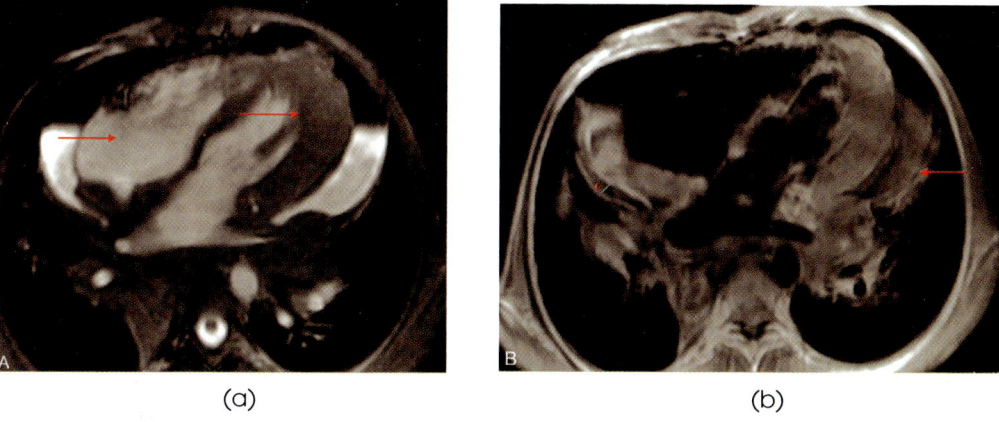

Fig. 11.4a and b: MRI cardiac showing large RA, hypertrophied LV measuring 2.5 cm (end diastolic thickness) and pericardial effusion (arrows)

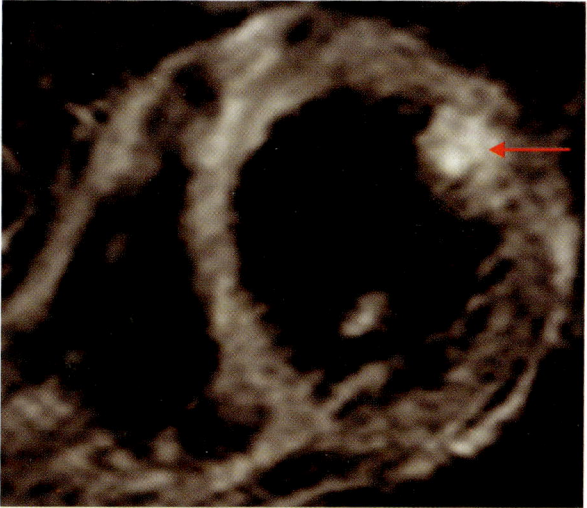

Fig. 11.5: Cardiac MRI showing heterogeneous enhancement of LV myocardium (arrow)

- Myocardial abnormalities are found in >50% of autopsied CSS patients.
- Cardiac manifestations comprise pericarditis, infiltrative cardiomyopathy, endomyocardial fibrosis, myocarditis, arrhythmias and sudden death.
- Symptomatic cardiomyopathy carries a poor prognosis, is responsible for nearly 50% of the deaths.

Blood Investigation Showed Gross Eosinophilia

Patient was kept on symptomatic treatment.

Conclusion

For CSS patients in remission, CMRI detected subclinical active myocardial lesions and could be recommended to assess cardiac involvement. However, because CMRI-detected MDE can reflect fibrosis or inflammation, FDG-PET might help to distinguish between the two.

Suggested Reading

1. Churg J, Strauss L. Allergic granulomatosis, allergic angiitis, and periarteritis nodosa. Am J Pathol 1951; 27: 277–301. Medline Web of Science Google Scholar.
2. McGavin CR, Marshall AJ, Lewis CT. Churg-Strauss syndrome with critical endomyocardial fibrosis: 10-year survival after combined surgical and medical management. Heart 2002; 87:E5.

Case 2

Patient's status is postoperative VSD closure and PDA ligation and non-union of sternal plates with overriding. Contrast chest CT thorax was advised.

Contrast CT Chest Findings

The sternum was separated with soft tissue between the two halves of the sternum suggesting features of sternal dehiscence (Fig. 11.6a and b).

Sternal wires were displaced to one side or the other for more than 2 cm with evidence of sternal wire breaks (Fig. 11.7a and b).

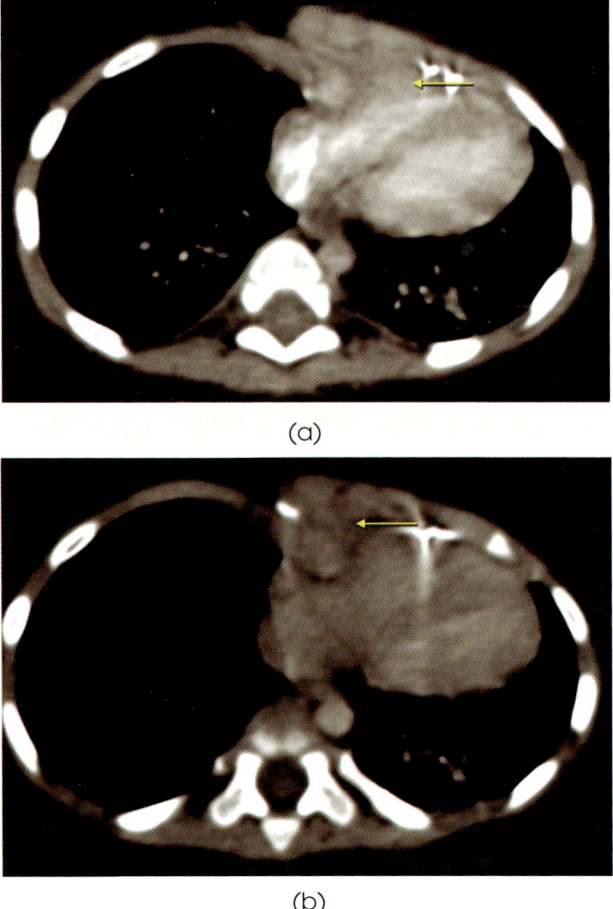

(a)

(b)

Fig. 11.6a and b: Contrast CT chest showing separated sternum with soft tissue between the two halves (yellow arrows)

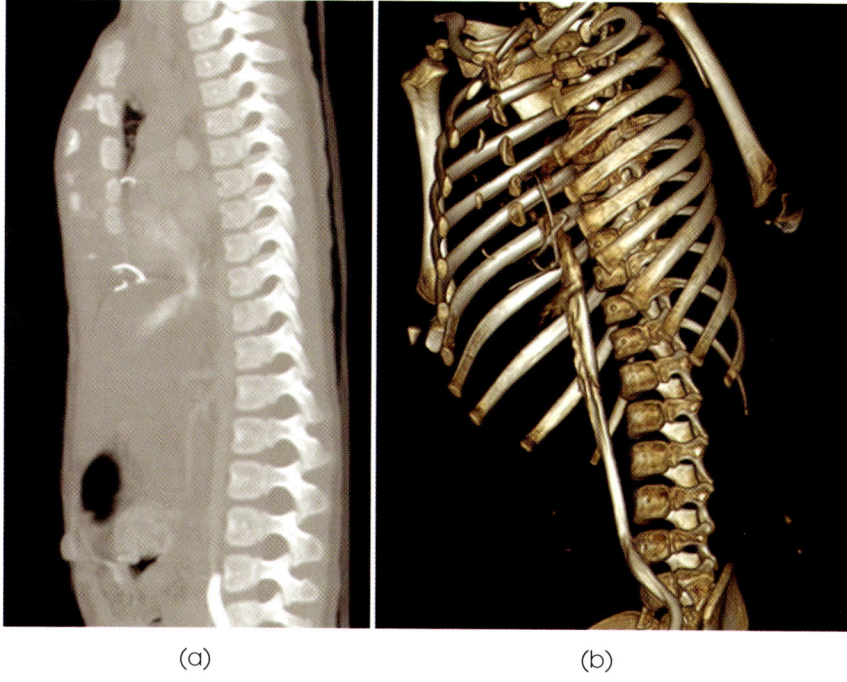

(a)   (b)

Fig. 11.7a and b: (a) Sagittal reformats and (b) Volume rendering image showing separation of sternal sutures and sternal dehiscence

Radiological Diagnosis

Sternal dehiscence

Surgical Intervention

Sternal refixation (Robicsek technique)

Discussion

Sternal dehiscence is an uncommon complication of median sternotomy occurring in approximately 3% to 5% of patients. It is potentially fatal and can be a difficult diagnosis clinically. Risk factors include prolonged ventilator support, blood loss requiring transfusion, postoperative wound infection, obesity, age, chronic obstructive pulmonary disease (COPD), and female gender. This complication usually occurs during the initial hospital stay.

The diagnosis is made radiographically in most cases. Plain X-ray of the sternum will most commonly reveal sternal wire abnormalities. Displacement of one or more wires is the most common finding. Other findings may be broken or rotated wire sutures. CT may be valuable in detecting the extent of any associated wound infection.

Treatment of sternal dehiscence is generally debridement and rewiring.

Conclusion

Sternal dehiscence is an uncommon complication that is most commonly diagnosed radiographically. The most common finding is displacement of a sternotomy wire.

Case 3

A 56-year-old female was referred to thoracic surgery department of the Max Superspeciality hospital, Delhi with complaints of gradually progressive dyspnea of 3 months duration. Clinically, the patient was hemodynamically stable. Routine blood investigations were within normal limits. A chest radiograph revealed large soft tissue opacities occupying mid and lower zones of both lung fields. The chest skiagram was not diagnostic. (Fig. 11.8).

A computed tomography (CT) scan of the thorax revealed a large mass measuring 10.5 cm × 8.4 cm × 6.5 cm of fat attenuation (−80 to −120 HU) with minimally enhancing internal densities (Fig. 11.9a and b). The mass was compressing the lungs bilaterally. There was cranial extension of mass in neck on right side and within the neck and superior mediastinum. The lesion shows a few enhancing soft tissue and admixed fat density (Fig. 11.10).

MRI of chest was done showed large fat intensity mediastinal lesion with a few fibrous strands in it. The lesion was extending in bilateral hemithorax causing compression and displacement of lungs resulting in reduction of air space (Fig. 11.11). The lesion had cranial extension into neck on right side and showed a few enhancing soft tissue admixed with fat intensity.

The patient underwent median sternotomy and a large lobulated fatty mass in the anterior mediastinum that was well encapsulated and was extending to surrounding recesses was noted and it was excised en bloc (Fig. 11.12). The mass weighed 15 kg.

The histopathological examination of the specimen showed a lesion composed of an admixture of mature adipose tissue and microscopically normal thymus tissue with Hassal's corpuscles, features that are consistent with thymolipoma.

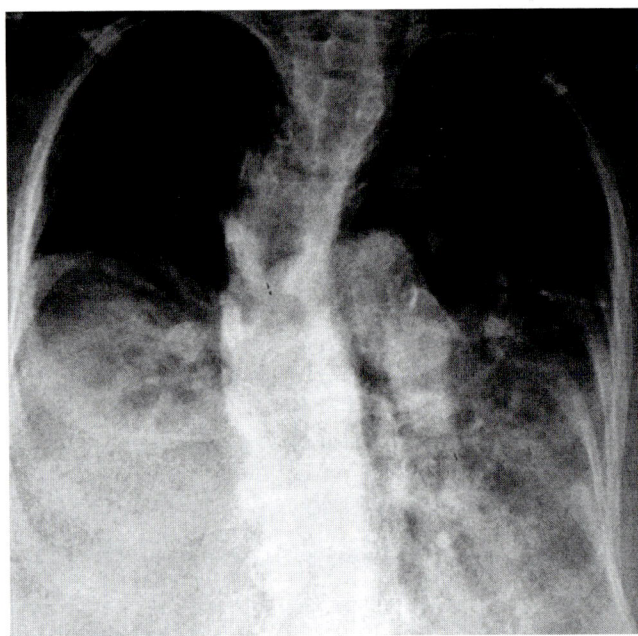

Fig. 11.8: Chest ski gram showing large soft tissue opacities occupying mid-lower zones of both lung fields

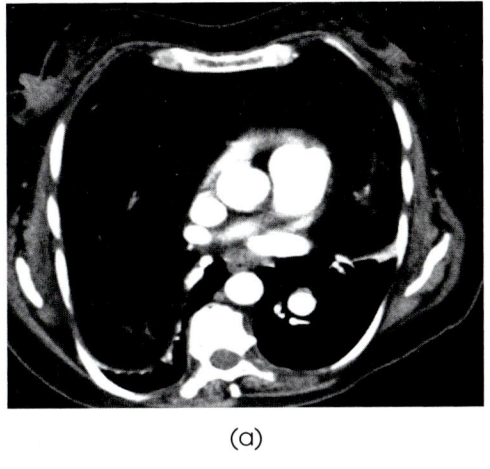

(a)

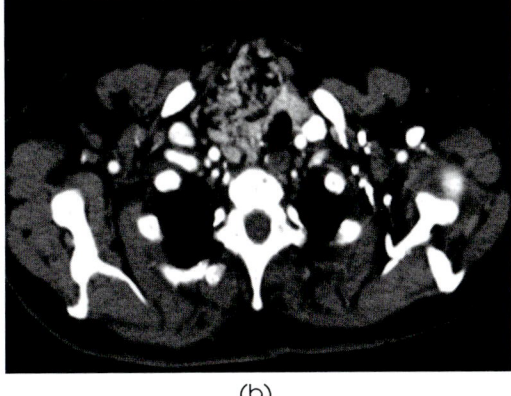

(b)

Fig. 11.9a and b: Chest CT showing large fat attenuating anterior mediastinal mass extending in bilateral hemithorax and neck region on right side

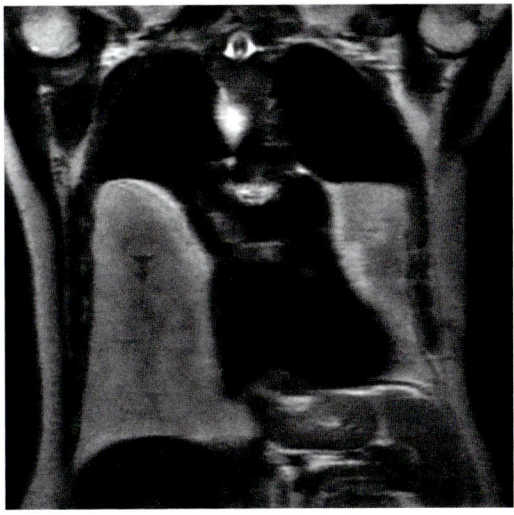

Fig. 11.10: MRI coronal T2W image showing large fat intensity mediastinal mass

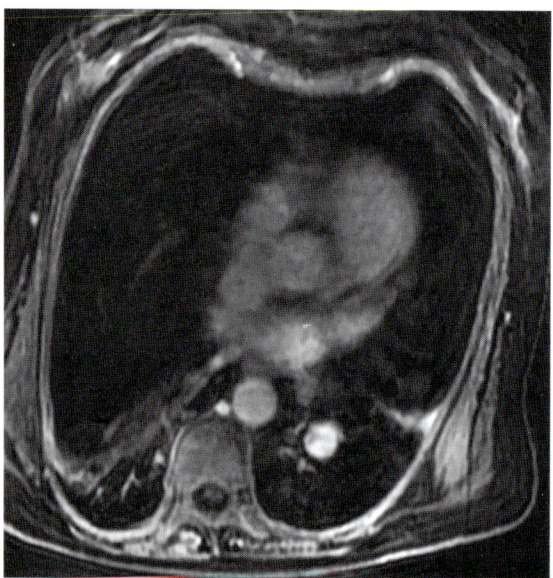

Fig. 11.11: Contrast MRI showing nonenhancing fat intensity mediastinal mass

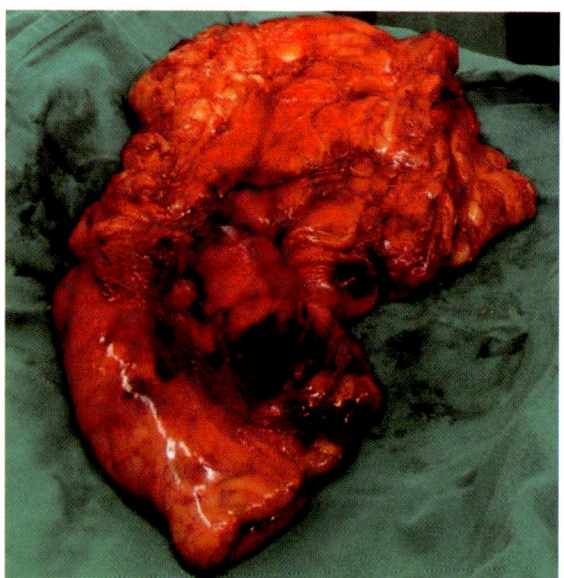

Fig. 11.12: Histopathological specimen showing the tumour en bloc

Conclusion

Thymolipoma is an uncommon benign neoplasm that accounts for 2% to 9% of thymic tumours. The weight of these tumours, according to our review of the literature, ranges from 154 g to 6000 g; in our case, the tumour weighed 15 kg.

This is a very rare tumour with a big size occupying more than 70% of intrathoaracic volume. There is no gender difference in its incidence, and the onset can occur at any age. It is also known as lipoma of the thymus, lipothymoma and thymolipomatous hamartoma, although we believe that thymolipoma is the most appropriate and clear designation. Thymolipoma may present as a symptomatic tumour that does not affect adjacent structures, and so large sizes can be reached before diagnosis. When the patient experiences symptoms, they are usually related to compression of adjacent structures, such as the heart, the great vessels, the lungs or the bronchi. Approximately half of the patients with this tumour experience symptoms such as dyspnea, chest pain, and weight loss, and myasthenia gravis is present in up to 10% of the cases. The weight of these tumours, according to our review of the literature, ranges from 154 g to 6000 g; in our case, the tumour weighed 15 kg. Characteristic CT findings involve fatty tissue containing soft tissue streaks that probably represent islands of normal thymic components. Fat density on CT scans, however, can also indicate lipoma, lip sarcoma, teratoma, prominent epicardial fat pad or congenital diaphragmatic hernias. In general, MRI scans show areas of high signal intensity intermingled with areas of intermediate intensity on T2-weighted images, and the adipose tissue of the tumour appears isointense in T1-weighted images. Although the diagnosis is strongly suggested by imaging studies that demonstrate fat and soft tissue within the tumour without the invasion of neighbouring structures, it is not possible to make an exact diagnosis or even to differentiate benign from malignant neoplasms. Although biopsy can be technically feasible, it is not indicated in these cases, because the mass must be surgically addressed and, if possible, completely excised.

The weight of these tumours, according to our review of the literature, ranges from 154 g to 6000 g; in our case, the tumour weighed 15 kg.

Suggested Reading

1. Roque C, Rodríguez P, Quintero C, Santana N, Hussein M, Freixinet J. Giant thymolipoma [Article in Spanish]. Arch Bronconeumol. 2005; 41(7): 402–3.
2. Shields TW, Robinson PG. Mesenchymal tumours of the mediastinum. In: Shields TW, LoCicero J, Ponn RB, editors. General Thoracic Surgery, Vol. 2. 5ed. Philadelphia: Lippincott Williams & Wilkins; 2000. pp. 2357–423.
3. Shirkhoda A, Chasen MH, Eftekhari F, Goldman AM, Decaro LF. MR imaging of mediastinal thymolipoma. J Comput Assist Tomogr. 1987; 11(2): 364–5.
4. Argani P, Rosai J. Thymoma arising with a thymolipoma. Histopathology. 1998; 32: 5568–78.
5. Wang Y, Sun Y, Zhang J. Diagnosis, treatment and prognosis of thymoma: an analysis 116 cases. Chin Med J. 2003; 116:1187–90.

Case 4

Patient was known case of right lung hydatid cyst. Patient had complaints of chest pain and breathlessness since 6 months.

Echocardiography was done which showed clot in left atrium. A cyst-like structure was also noted in left atrium. Cardiac MRI was advised for further evaluation.

Chest CT and Cardiac MRI Findings (Figs 11.13 and 11.14)

There was large cyst occupying lower lobe of right lung.

There was dilated left atrium with severe mitral stenosis. A hypointense non-enhancing lesion was seen in left atrial appendage likely clot. Rest of the chambers were normal in size.

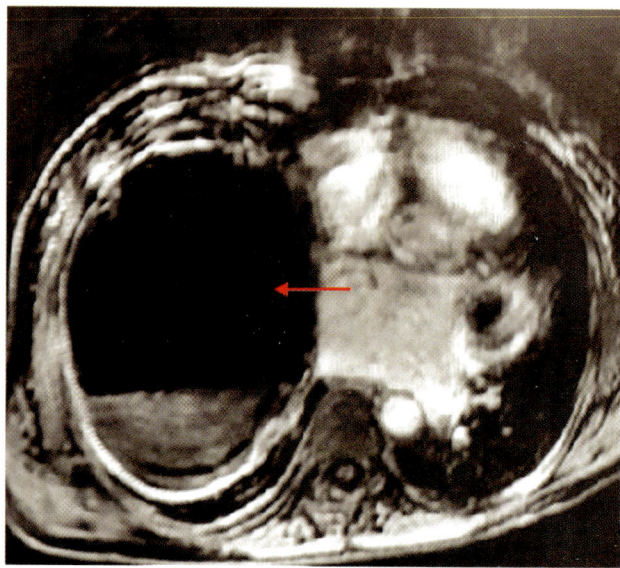

Fig. 11.13: Showing right lung hydatid cyst (arrow)

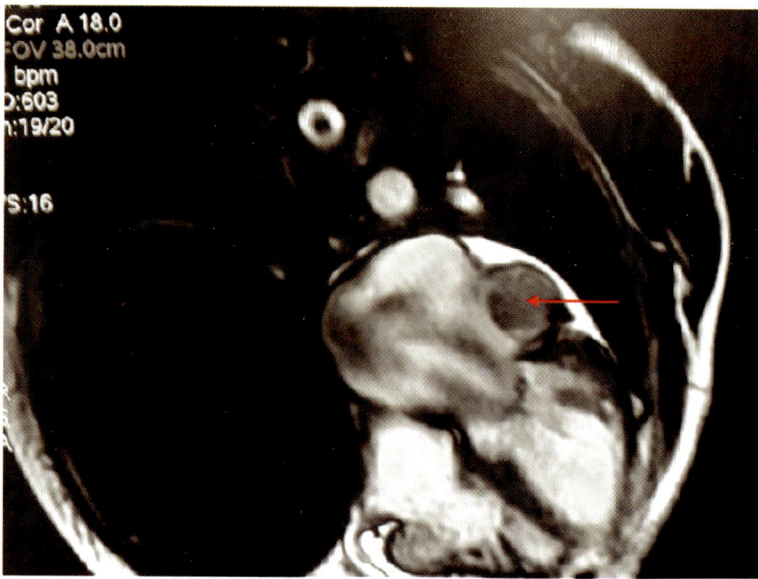

Fig. 11.14: Showing clot in left atrial appendage (arrow)

Surgical Intervention

Hydatid cyst was resected out. Mitral valve replacement and left atrial appendage clot removal was done.

Conclusion

Hydatid disease is a manifestation of larval infestation by the echinococcus tapeworm, common in various endemic regions. In adults, the lungs are second-most common organ to be involved by hematogenous dissemination. Uncomplicated pulmonary hydatid cysts are most commonly diagnosed incidentally on imaging. Though a variety of signs has been described on imaging, complicated hydatid cysts can present with atypical imaging findings. CT is the imaging modality of choice, especially in complicated hydatid cysts, and can provide an accurate diagnosis by demonstrating the internal characteristics and morphology of the lesion. Thus, radiologists should be well aware of the typical and atypical imaging features of the disease. In our case MRI was helpful in diagnosing the clot in left atrium.

Suggested Reading

1. Beggs I. The radiology of hydatid disease. AJR Am J Roentgenol. 1985; 145: 639–648.
2. Lewall DB. Hydatid disease: biology, pathology, imaging and classification. Clin Radiol. 1998; 53: 863–874.

Case 5

Patient is 5-year-old male of known case of glucose-6-phosphate dehydrogenase (G6PD) deficiency. Patient has lower neck and mediastinal mass had surgery once and now present with recurrence. (Histopath proven mediastinal desmoid tumour. CECT chest and MRI chest were done in Max Hospital for further evaluation.)

CT Chest Findings (Figs 11.15, 11.16a, b, and 11.17)

Well-defined homogenous hypodense mass lesion extending from true cord level compressing thyroid lobes extending caudally encasing right brachiocephalic artery, bilateral carotid vessels, and large area of contact with arch of aorta and further extending in anterior mediastinum.

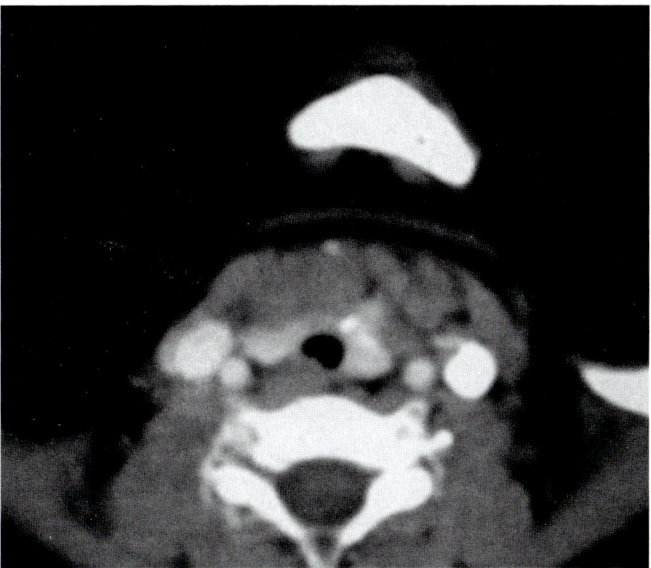

Fig. 11.15: CECT chest showing well-defined homogeneous hypodense mass lesion extending from true cord level compressing thyroid lobes

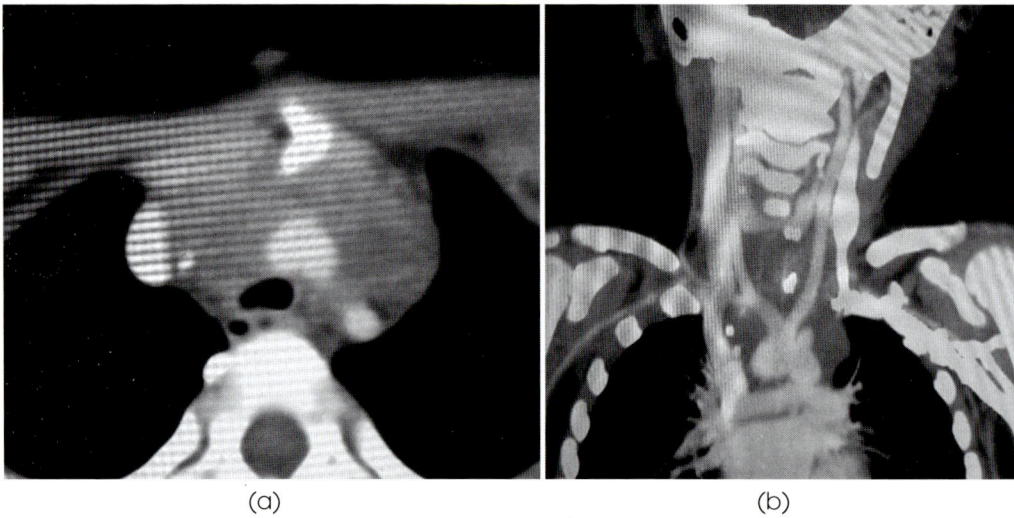

(a)   (b)

Fig. 11.16a and b: CECT chest showing hypodense mass encasing RT brachiocephalic art, B/L carotid vessels and encasing right brachiocephalic art, B/L carotid vessels

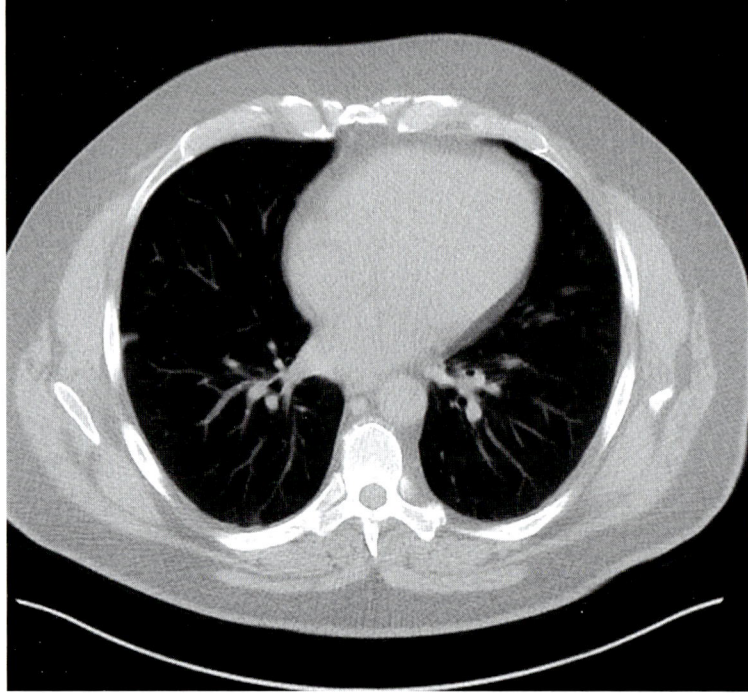

Fig. 11.17: HRCT chest showing normal lung parenchyma

Contrast MRI Chest (Figs 11.18, 11.19 and 11.20)

Was done which showed consistent findings, however, MRI delineation of the soft tissue details were better due to better temporal and spatial resolution and multiplanar abilities.

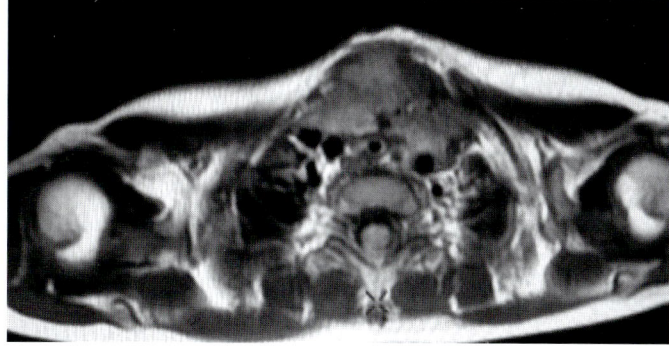

Fig. 11.18: MRI chest

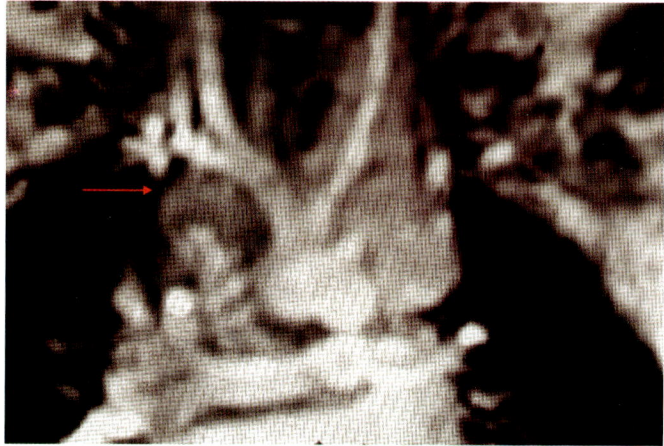

Fig. 11.19: MRI chest showing mass encasing the great vessels however, there is no invasion of any of the vessel (arrow)

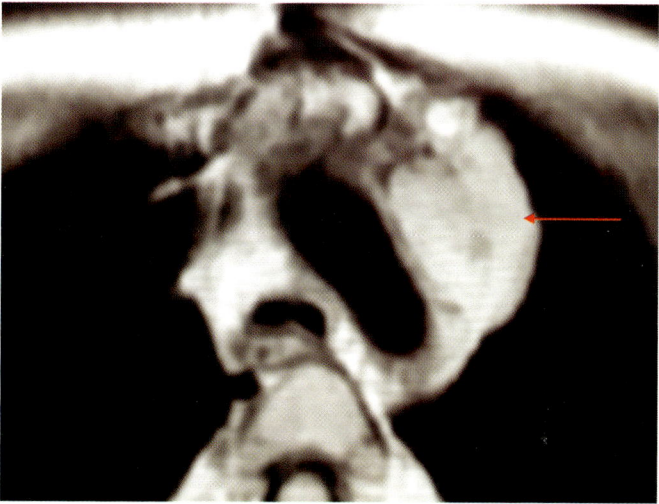

Fig. 11.20: Chest MRI showing mass in approximation to arch of aorta (arrow)

Diagnosis

Large mediastinal mass likely desmoid tumour. Biopsy was done which confirmed it.

Medical Management

Patient was given chemotherapy and follow-up MRI showed significant resolution of the mass.

Case 6

Patient has complaints of breathlessness off and on. Echocardiography was done showed a mass having heterogeneous echoes lying at posterior aspect of left atrium causing extrinsic compression over posterior wall of left atrium.

Cardiac MRI was done to rule out this extrinsic mass causing compression over posterior wall of left atrium.

Cardiac MRI Findings

Large hiatal hernia with stomach and surrounding mesentric fat herniated in posterior mediastinum compressing the post wall of left atrium. No evidence of any other mass in posterior mediastinum or lying in proximity of left atrium. The hiatal herna itself compressing the left atrium (Figs 11.21 and 11.22a).

Radiological Diagnosis

Hiatal hernia compressing posterior wall of LA (Fig. 11.22b).

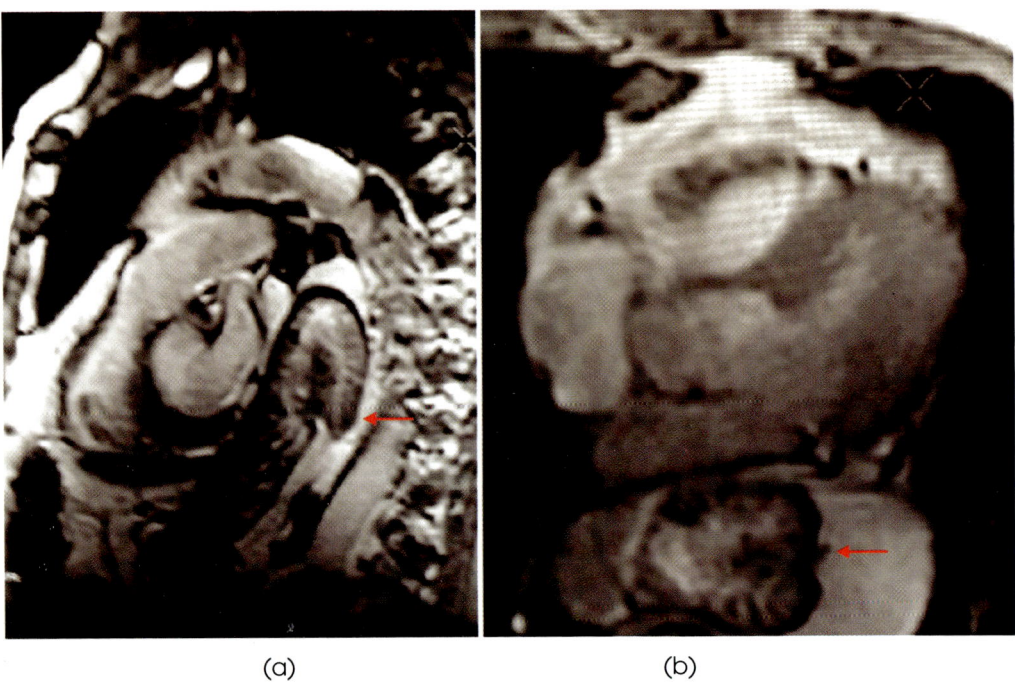

(a)   (b)

Fig. 11.21: Cardiac MRI showing stomach and mesentric fat herniating in posterior mediastinum (arrow)

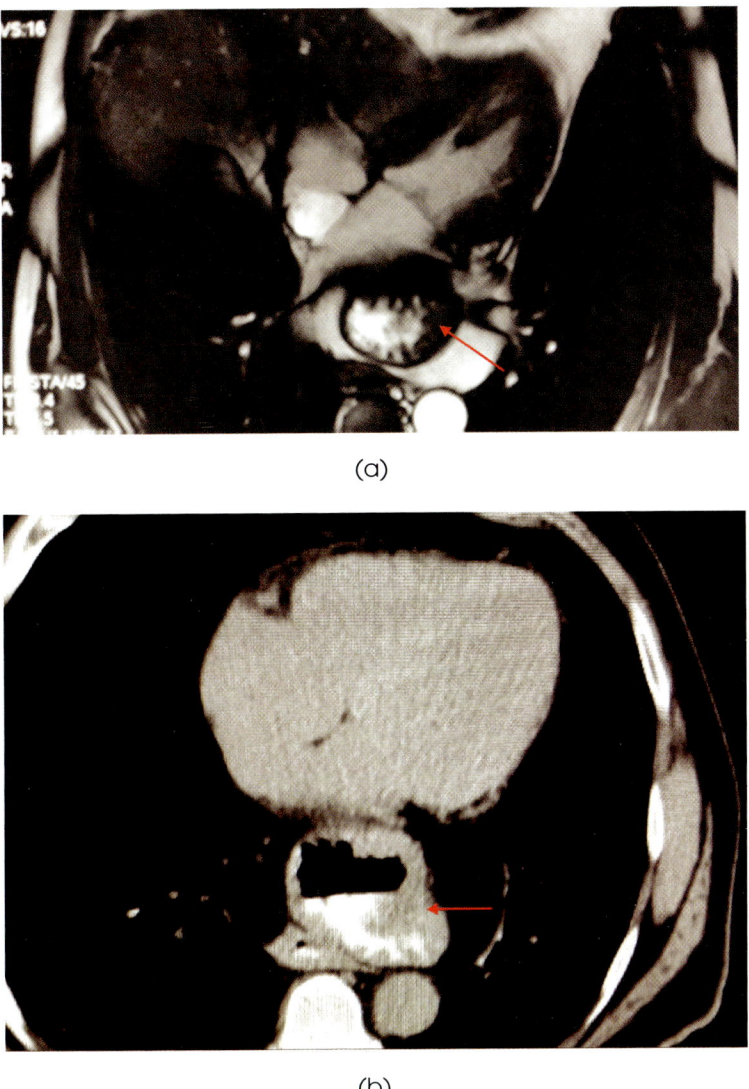

(a)

(b)

Fig. 11.22a and b: Cardiac MRI 4-chamber and CT scan image showing hiatal hernia. Oral contrast in stomach (arrow)

Patient has c/o of breathlessness and chest pain off and on.
CT scan and MRI were done for further evaluation.

CT Scan and MRI Chest Findings

An approximately 8 × 6 × 6 cm anterior medistinal mass lying extrapericardially on the right side, abutting the right lung on its lateral aspect and abutting RA and SVC medially. The mass was enhancing heterogenously and showed fat component and calcification. Thymus gland was normal and no enlarged lymph nodes were seen (Figs 11.23a, b 11.24, 11.25a and b).

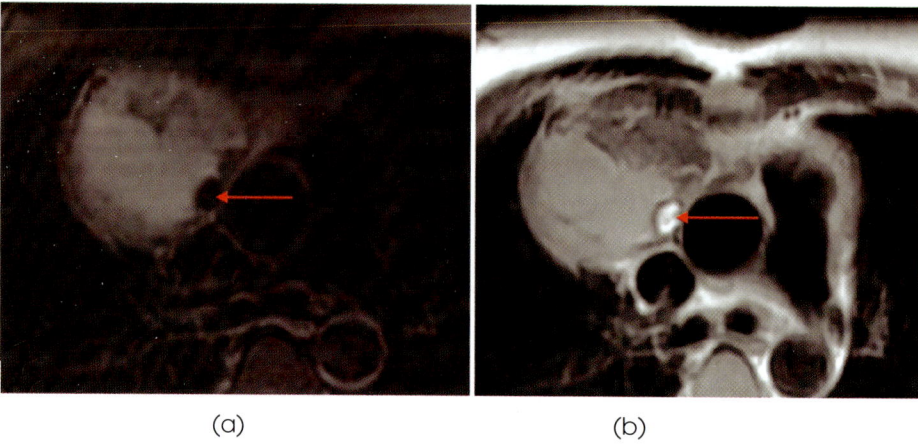

(a)   (b)

Fig. 11.23a and b: MRI showing fat content in the anterior mediastinal mass (arrow)

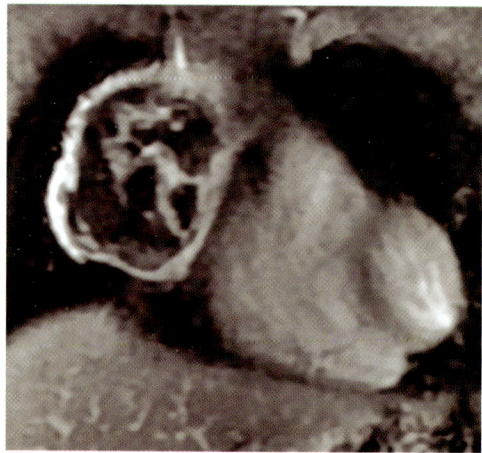

Fig. 11.24: MRI showing mediastinal mass abutting pericardium, enhancing heterogeneously with enhancing capsule.

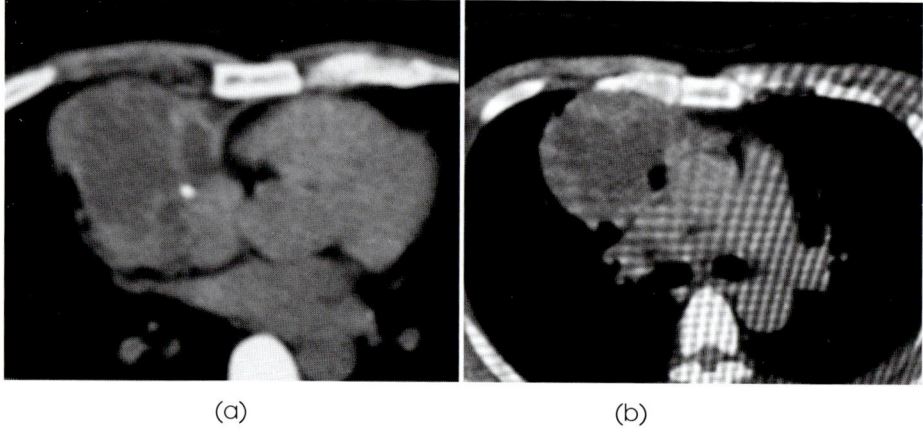

(a)   (b)

Fig. 11.25a and b: CT showing calcification, necrosis and fat component

Radiological Diagnosis

- Benign cystic teratoma
- D/D thymic lesions

Surgical Intervention

Surgical removal of the mass was done (Figs 11.26 and 11.27).

Anterior mediastinal mass—mature cystic teratoma.

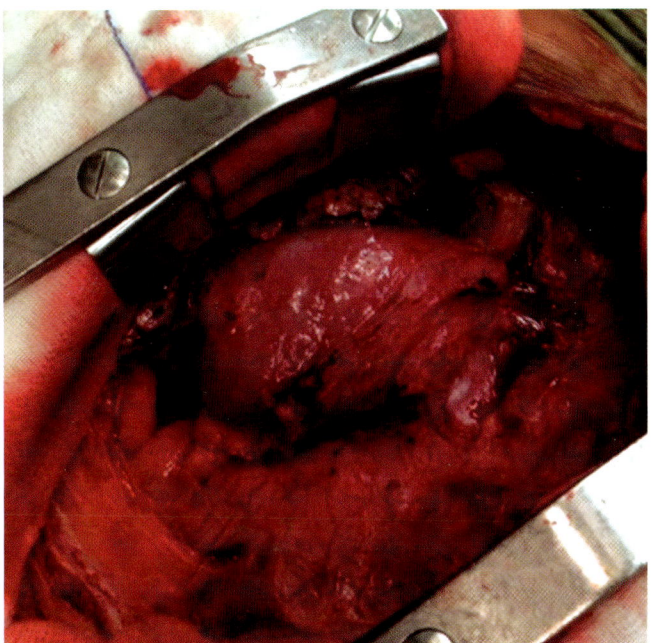

Fig. 11.26: Surgical sample

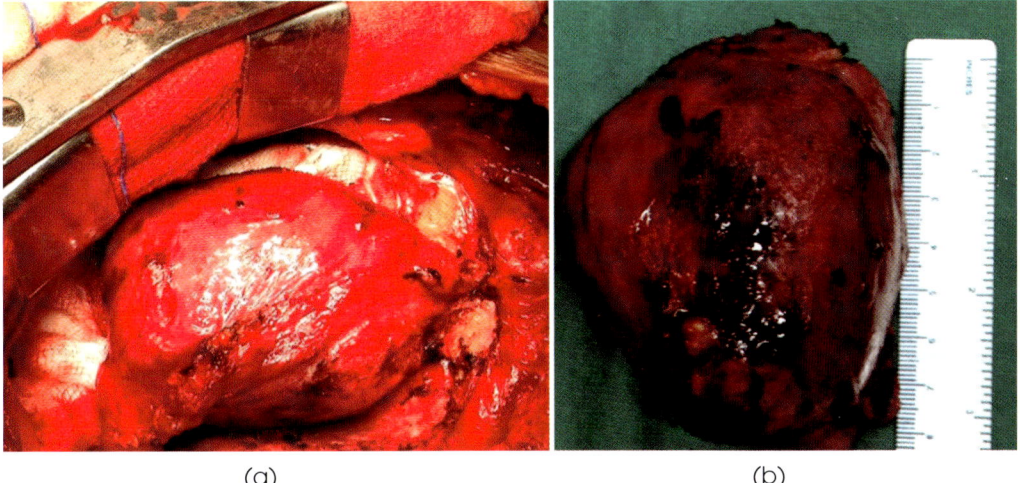

(a) (b)

Fig. 11.27: Surgical sample

Conclusion

Anterior mediastinal masses comprise a wide spectrum of tumours. Multimodality imaging is important in characterizing and staging of these tumours. Although conventional CT remains the modality of choice, cardiac-gated MRI has an increasingly important role, especially in tissue characterization and assessment of surgical resectibility.

Suggested Reading

1. Priola SM, Priola AM, Cardinale L, Perotto F, and Fava C. The anterior mediastinum: anatomy and imaging procedures. Radiol Med. 2006; 111: 295–311.
2. Whitten CR, Khan S, Munneke GJ, Grubnic S. A diagnostic approach to mediastinal abnormalities. Radiographics. 2007; 27: 657–671.

Vascular Anomalies and Diseases

Case 1

A 4-month-old baby came with complaints of breathlessness. Echocardiography from outside showed VSD and suspicion of coarctation of aorta.

Cardiac MRI was done.

The Imaging Findings Were

1. Large VSD (Figs 12.1 and 12.2)
2. Dilated pulmonary artery (Fig. 12.3)
3. Hypoplasia of aorta with coarctation (Figs 12.4, 12.5a and b)
4. High pressure gradient across coarctation as measured by cardiac MRI venc study.

Surgical Intervention

Ballooning of coarctation segment and VSD closure were done.

Discussion

Aortic coarctation is one of the most common congenital heart diseases, and early diagnosis and treatment are the keys to successful outcomes. Cardiac echocardiography is generally the first imaging test because of its ease of use and lack of ionizing radiation.

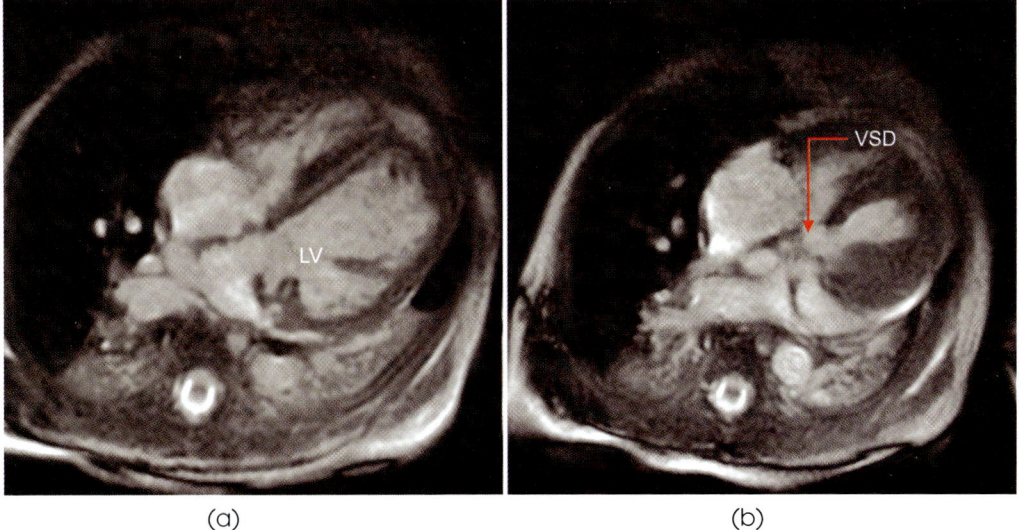

(a)   (b)

Fig. 12.1: MRI 4CH view showing VSD, dilated LV

However, not all segments of the aorta can be optimally evaluated with this modality and CT or MRI is almost always used for surgical planning and follow-up.

With the use of these cross-sectional modalities, the surgical and medical care teams can be provided with accurate information for surgical planning and follow-up in a completely noninvasive manner. Particularly with the functional information provided with phase contrast imaging, the pressure gradient across the coarctation segment, which is an important indicator during the decision-making process, can be accurately predicted. Another important advantage of these noninvasive imaging modalities is simultaneous evaluation of the heart for associated congenital defects.

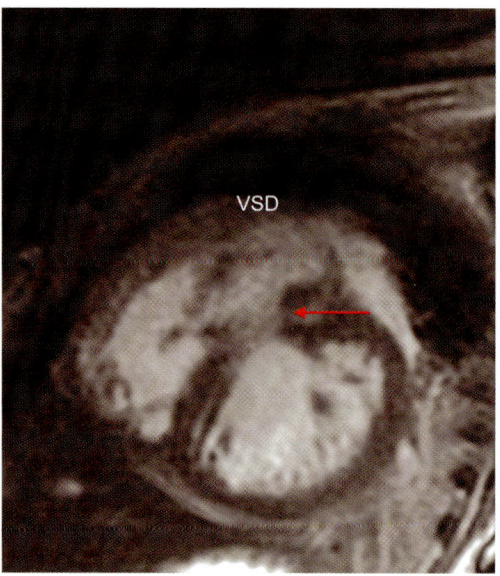

Fig. 12.2: MRI short axis image showing VSD (arrow)

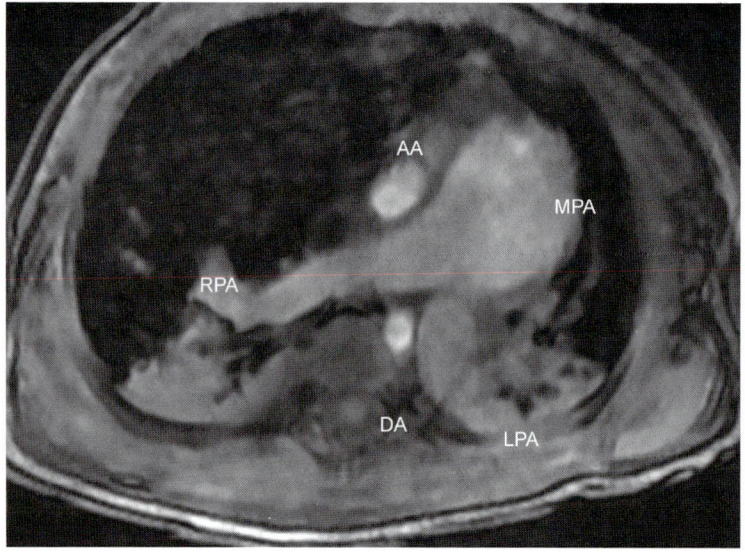

Fig. 12.3: MRI showing dilated pulmonary artery. MPA—25.3 mm, RPA—8.8 mm, LPA—11.1 mm

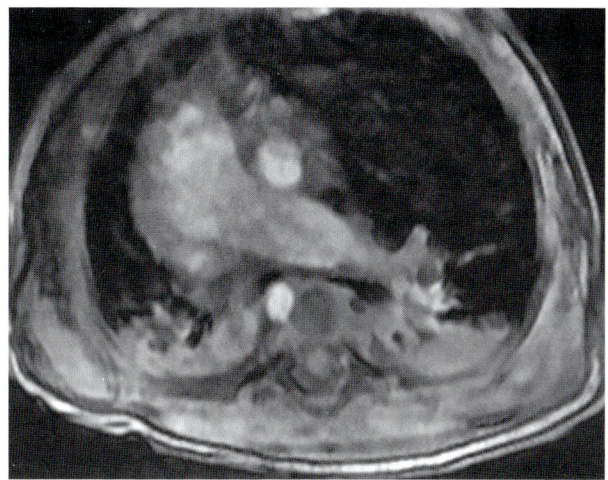

Fig. 12.4: MRI showing hypoplasia of aorta including ascending, arch and descending aorta. Ascending aorta—10.8 mm. Arch of aorta—6.3. Descending aorta—6 mm. At diaphragm —4 mm.

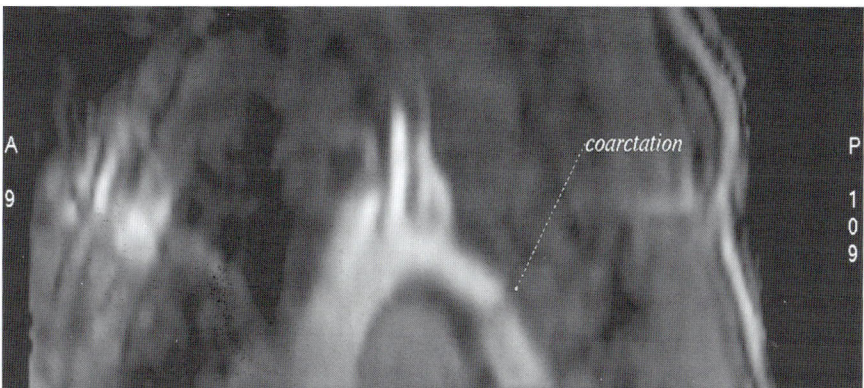

Fig. 12.5a: Stenotic segment—3.8 mm in diameter, length—5 mm, prestenotic segment— 5.9 mm, poststenotic segment—6.9 mm, stenotic segment at distance of 2 cm from origin of subclavian artery. Pressure gradient—20 mm Hg

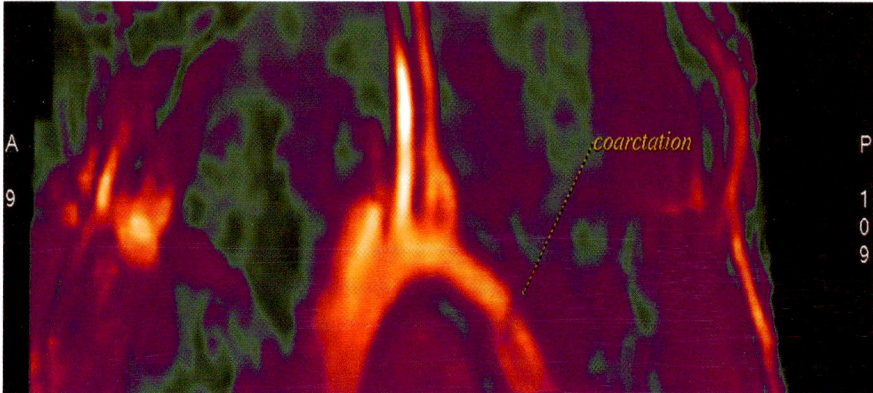

Fig. 12.5b: Reconstructed image

Suggested Reading

Kimura-Hayama ET, Meléndez G, Mendizábal AL, Meave-González A, Zambrana GF, Corona-Villalobos CP. Uncommon congenital and acquired aortic diseases: role of multidetector CT angiography. RadioGraphics 2010; 30: 79–98.

Case 2

Patient is 65-year-old female came with complaints of chronic left upper limb ischemia.

There was gangrenous fingertips of left hand and flexure contracture of left elbow. There was minimal movement of left wrist. Patient was referred for CT angiography in radiology department.

CT Angiography Findings

There was circumferential soft tissue plaque involving origin and proximal portion of left subclavian artery causing 80% stenosis (Fig. 12.6a and b) and soft plaque in mid-left brachial artery causing 80–90% stenosis (Fig. 12.7a and b). Distal left brachial artery was showing contrast opacification as was supplied by collaterals.

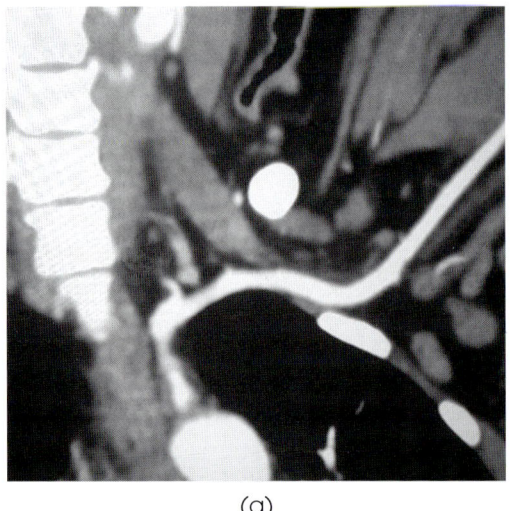

(a)

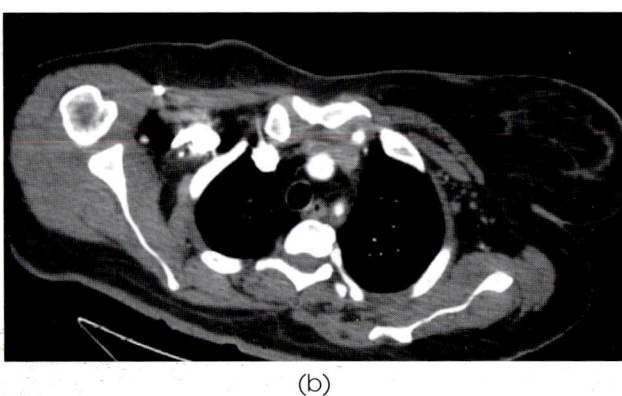

(b)

Fig. 12.6a and b: CT angiography showing circumferential soft tissue plaque involving origin and proximal portion of left subclavian causing 80% stenosis

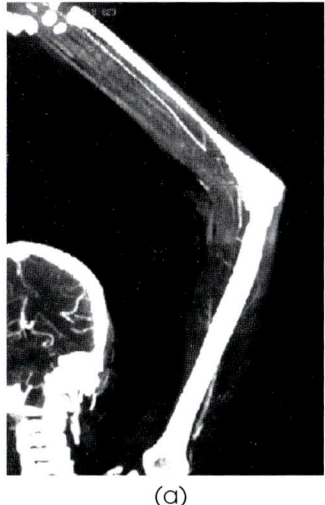
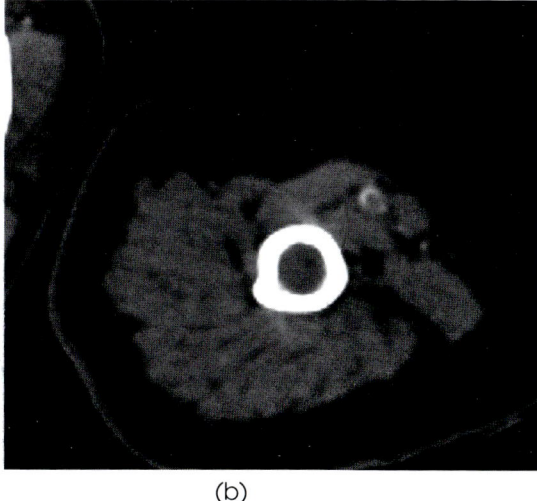

(a)   (b)

Fig. 12.7a and b: CT angiography showing soft plaque in mid-left brachial artery causing 80–90% stenosis

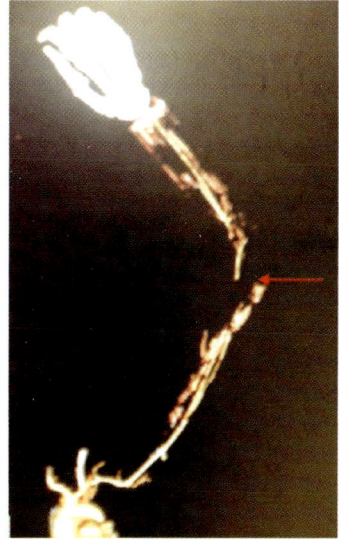

Fig. 12.8: Volume rendering image showing the stenosed segment of brachial artery

Surgical Intervention

Left subclavian angioplasty was done on 20/10/14 and patient was discharged on 21/10/14.

Conclusion

While there are direct and indirect benefits of diagnosing subclavian stenosis, it is a disease entity that is often overlooked in our patient population. Evidence exists to support the use of non-invasive screening tools with excellent negative predictive capabilities. Therefore, it is important that the vascular medicine practitioner be aware of the disease process, symptoms and exam findings, as well as treatment options for their patients. Medical therapy for atherosclerotic risk factors should be standard care for all patients with any degree of subclavian stenosis. When an intervention is indicated, a percutaneous approach is favored over surgical intervention in the current era of angioplasty and stenting.

Suggested Reading

1. English JA, Carell ES, Guidera SA, Tripp HF. Angiographic prevalence and clinical predictors of left subclavian stenosis in patients undergoing diagnostic cardiac catheterization. Catheter CardiovascInterv 2001; 54: 8–11.
2. Gutierrez GR, Mahrer P, Aharonian V, Mansukhani P, Bruss J. Prevalence of subclavian artery stenosis in patients with peripheral vascular disease. Angiology 2001; 52: 189–194.

Case 3

Patient is follow-up case of atherosclerosis disease with redo right aorto-femoral popliteal graft and left aorto-femoral graft. Embolectomy and graft thrombectomy was done in 2013. Now patient is complaining of severe limiting claudication in left lower limb for 2 months. Patient also had history of diabetes/hypertension and CAD. Patient was referred for CT angiography.

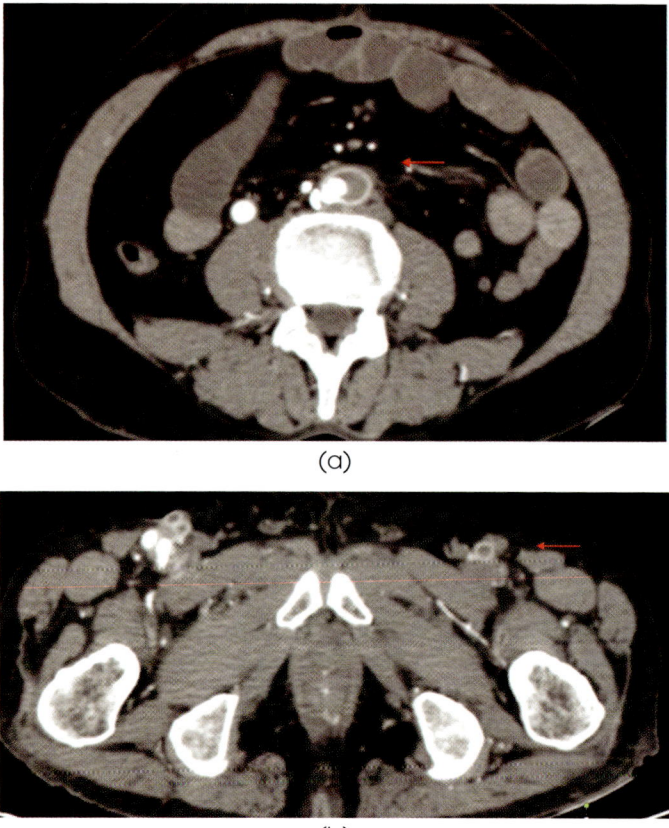

(a)

(b)

Fig. 12.9a and b: Showing blocked left aorto-femoral graft and patent right redo aorto-femoral graft (arrow)

CT Angiography Findings

There was diffuse atherosclerotic disease involving bilateral lower limb arteries with patent right aorto-femoral graft and occluded left aorto-femoral graft. There were multiple collaterals filling the arteries below left knee (Figs 12.9a, b, 12.10a, b, 12.11a, b and c).

Surgical Intervention

Peripheral bypass graft left femoral cross over bypass using PTF with graft.

Conclusion

CT angiography is well suited for the morphologic assessment of peripheral arterial bypass grafts. Compared with duplex US, multi-detector row CT angiography allows an accurate depiction of graft-related complications, including stenosis, occlusion, aneurysmal changes, and arteriovenous fistulas. When compared with duplex US and

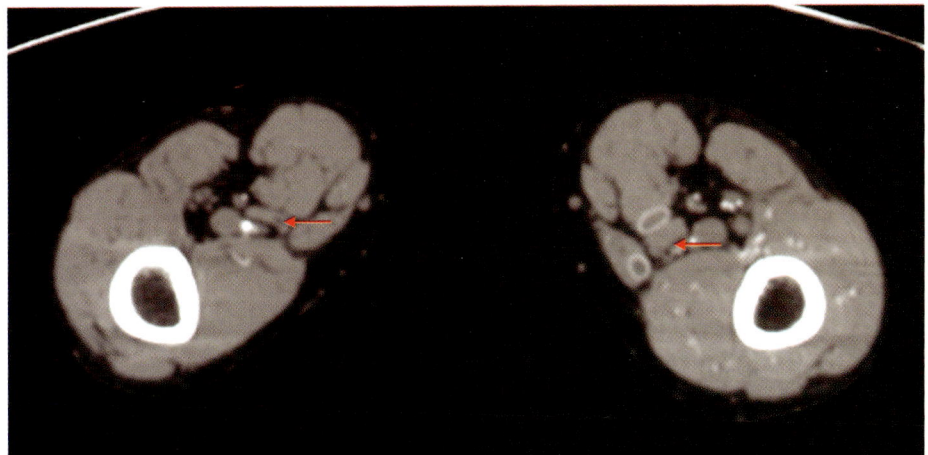

(a)

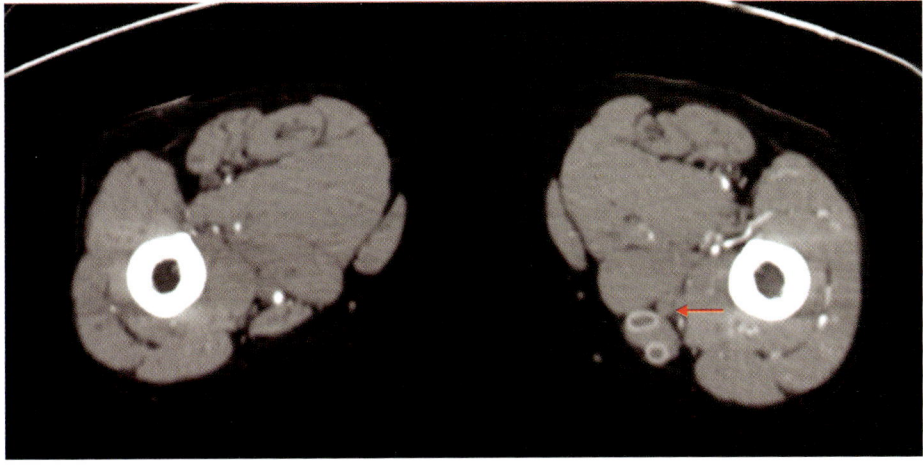

(b)

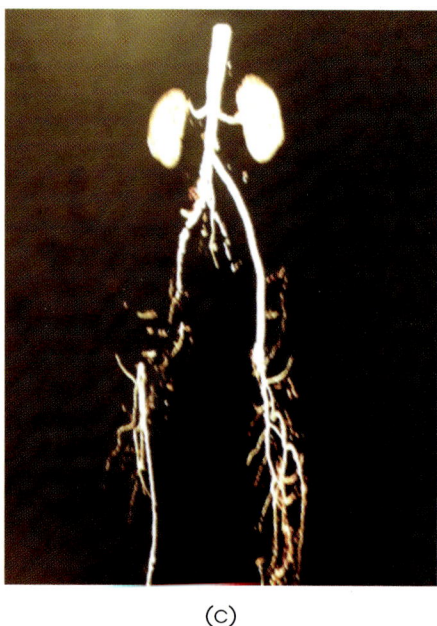

(c)

Fig. 12.10a to c: Volume rendering image showing blocked left aorto-femoral graft

conventional DSA, sensitivity and specificity values of more than 95% were achieved by both readers with multi-detector row CT angiography for the diagnosis of arterial bypass graft-related complications. In conclusion, this study has demonstrated in a prospective blinded comparison that multi-detector row CT angiography is feasible, accurate, and reliable in the assessment of peripheral arterial bypass grafts and detection of graft-related complications, including stenosis, aneurysmal changes, and arteriovenous fistulas. Because of its noninvasive nature and lower effective dose, multi-detector row CT angiography may replace conventional DSA as a technique to be used after performance of duplex US to help physicians plan further treatment of peripheral arterial bypass grafts.

Suggested Reading

1. Rubin GD, Shiau MC, Leung AN, Kee ST, Logan LJ, Sofilos MC. Aorta and iliac arteries: single versus multiple detector-row helical CT angiography. Radiology 2000; 215: 670–676.
2. Rubin GD, Schmidt AJ, Logan LJ, Sofilos MC. Multi-detector row CT angiography of lower extremity arterial inflow and runoff: initial experience. Radiology 2001; 221: 146–158.

Case 4

Patient is 87-year-old with history of swelling in right axilla. Initially small then increased significantly in size. Now it is associated with pain. Paresthesia in right upper limb. No history of trauma. No history of DM/HTN/CAD.

On examination: Large pulsatile swelling in right axillary region. Distal pulses were palpable.

CT Angiography Findings

Aneurysmal dilatation of right axillary artery involving mid-part about 3 cm from the origin. It measures 8 × 8 cm in size. Peripheral hypoattenuation is seen suggestive of thrombus formation. Right axillary and subclavian artery were ectatic and tortous (Figs 12.11a, b, 12.12 and 12.13).

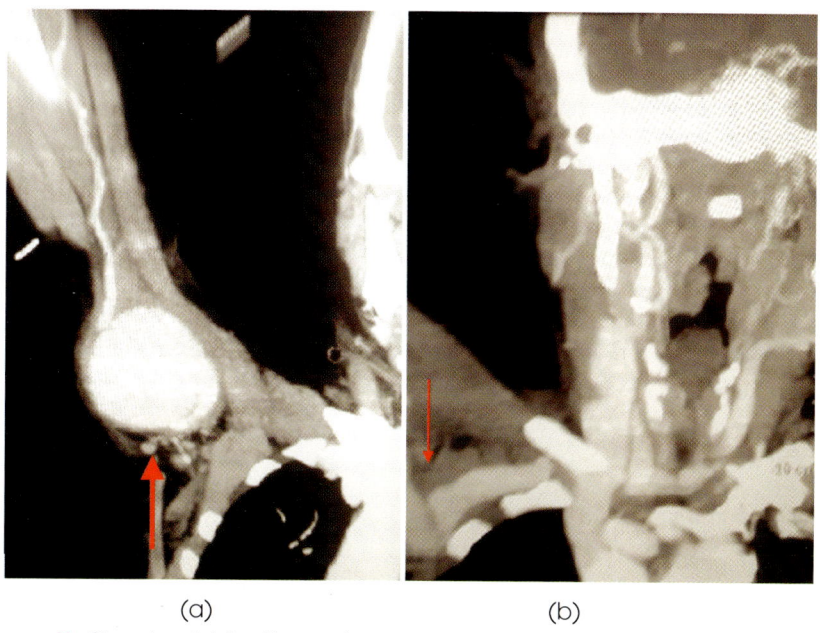

(a)   (b)

Fig. 12.11a and b: Showing right axillary artery aneurysm (arrow) and ectasia of right subclavian artery (thin arrow)

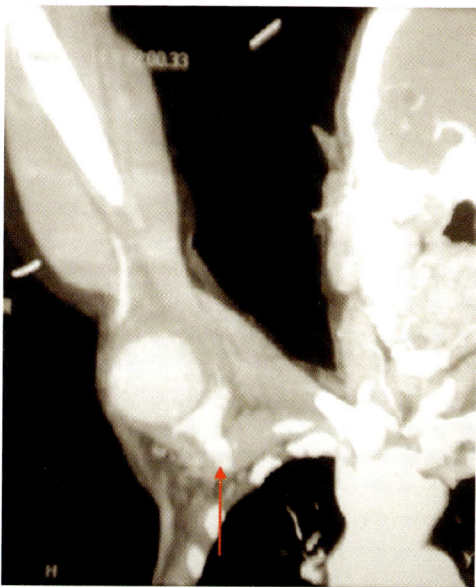

Fig. 12.12: Proximal to aneurysm

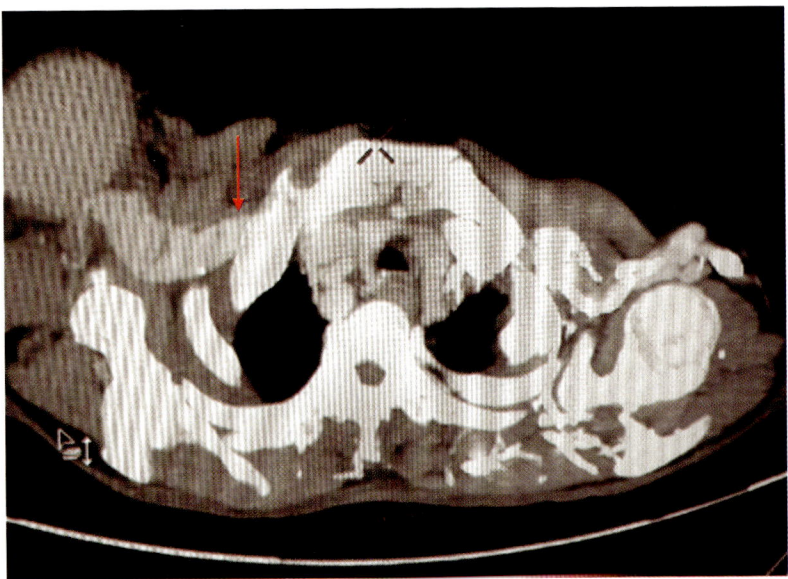

Fig. 12.13: Showing tortuous right subclavian artery (arrow)

Surgical Intervention

Procedure: Excision of right axillary artery aneurysm with subclavian artery to brachial bypass using PTFE interring graft 80 × 40 cm under GA done on 13/11/14. Patient was discharged on 15/11/14.

Conclusion

Aneurysms of the axillary artery are extremely rare in the literature. There is a report of conventional surgical treatment of an axillary artery aneurysm secondary to thoracic outlet syndrome. There are also reports of two cases of young patients, one 21 and the other 22, with congenital aneurysms of the subclavian artery. In summary, aneurysms of the subclavian and axillary arteries are rare and when found surgery is indicated in order to prevent complications. The approach should be chosen on a case-by-case basis, depending on the topography of the aneurysm and the patient's characteristics.

Suggested Reading

1. Davidović LB, Marković DM, Pejkić SD, Kovacević NS, Colić MM, Dorić PM. Subclavian Artery Aneurysms. Asian J Surg. 2003; 26(1): 7–11.
2. Akar H, Sarac A, Iriz E, Kolbakir F, Dermican S. Intrathoracic Aneurysm of the Right Subclavian Artery. ActaChir Belg. 2004; 104: 730–2.

Case 5

Patient is 45 years old female with history of old Koch's taken ATT now came with complaints of breathlessness on exertion for one month. Patient is known case of chronic renal disease (CKD) now has pulmonary hypertension. D-Dimer was normal. As contrast CT could not be done to rule out pulmonary embolism due to CKD status of the patient study. So noncontrast MRI cardiac and chest were advised to rule out pulmonary embolism and further evaluation of pulmonary hypertension. Pulmonary flow assessment was also done to see pressure gradient across main, right and left pulmonary arteries by VENC cardiac MRI. Left and right ventricular (LV, RV) qualitative analysis was also done. These all could be possible with MRI without using intravenous contrast.

Non-contrast Cardiac and Chest MRI Findings

- Dilated main, right and left pulmonary arteries (Figs 12.14a, b and 12.15).
- There was increased pressure gradient across pulmonary arteries as calculated by software in built in the MRI workstation using velocity encoding study (Fig. 12.16a, b and c).
- No obvious filling defect suggesting pulmonary embolism was noted.
- LV qualitative analysis revealed normal LV systolic function. Regional wall motion analysis showed normal motion in all LV and RV walls.
- Dilated RV and reduced RV systolic function (Fig. 12.17).

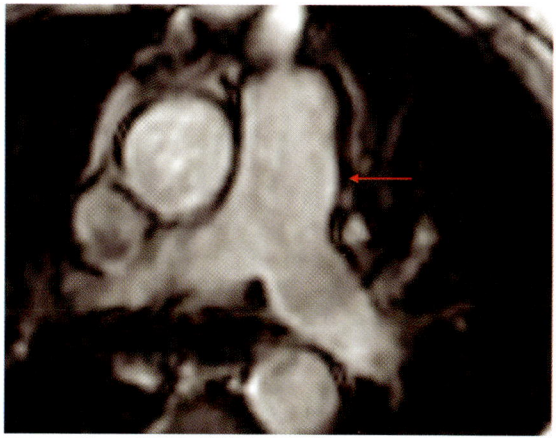

Fig. 12.14a: Chest MRI showing dilated MPA (arrow), RPA and LPA

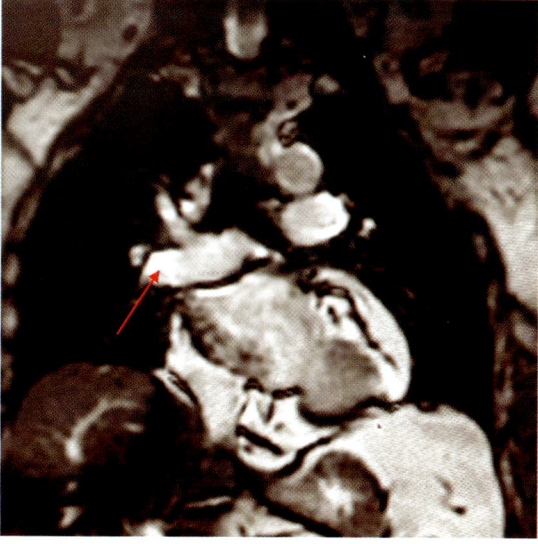

Fig. 12.14b: Cardiac MRI showing RPA and its branches (arrow)

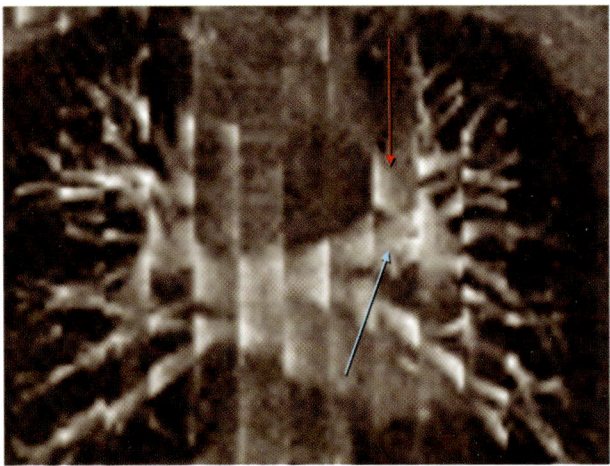

Fig. 12.15: Cardiac MRI pulmonary tricks showing RPA, LPA, their superior and inferior trunks and their ascending and descending branches. No evidence of filling defects suggestive thrombus seen.

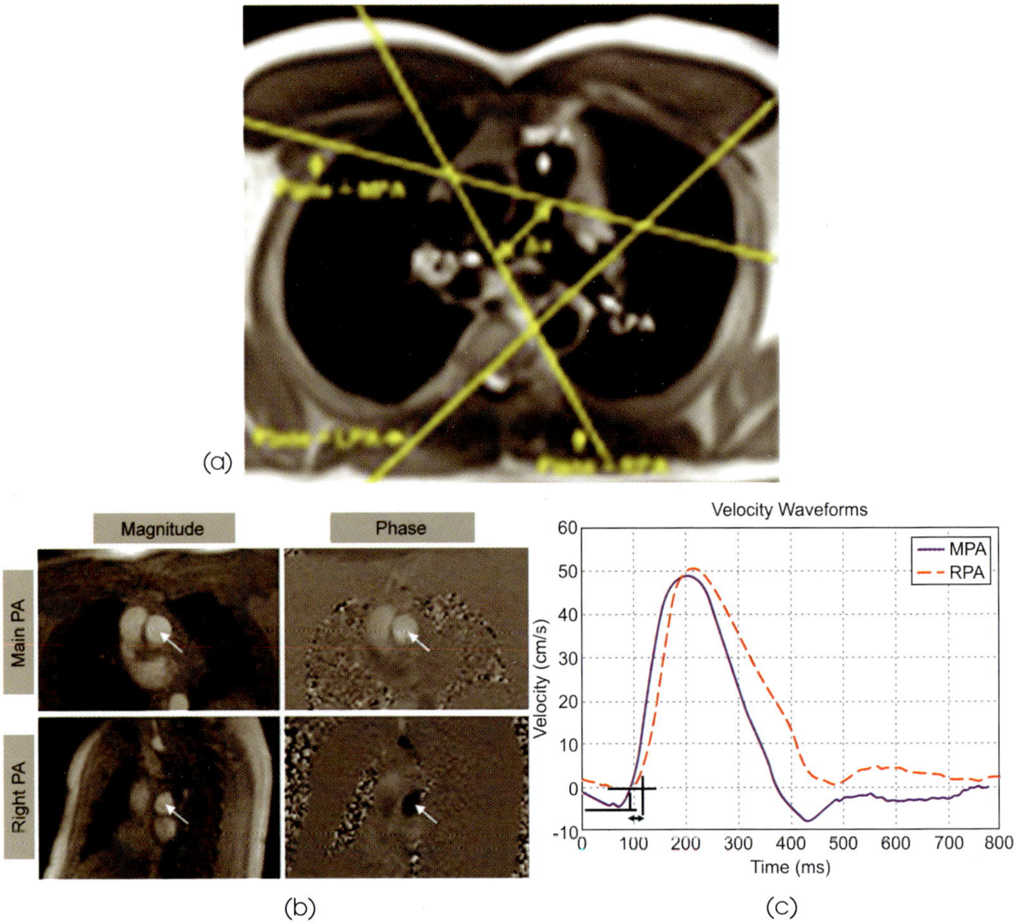

(a)

(b)   (c)

Fig. 12.16: Venc study for pulmonary artery gradients

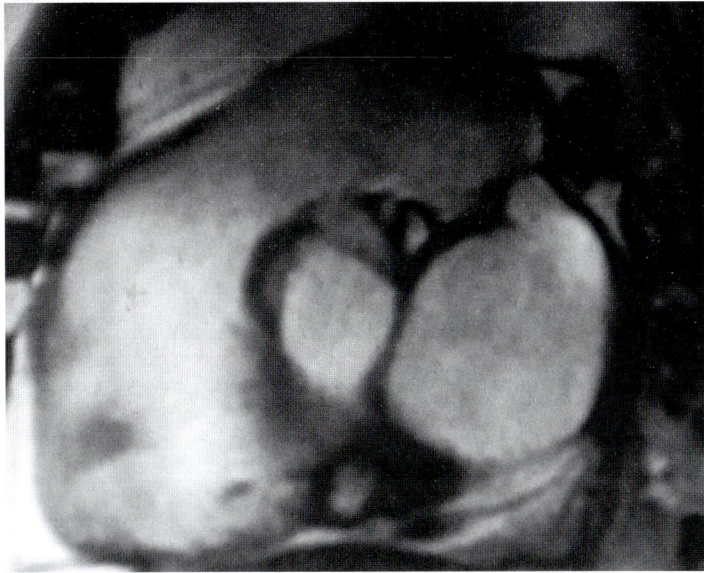

Fig. 12.17: cardiac MRI RVOT showing dilated right ventricle

Diagnosis

Dilated RV and reduced RV systolic function with increased pressure gradient across MPA. No evidence of pulmonary embolism noted.

Medical management

In absence of pulmonary embolism patient was discharged under treatment of pulmonary hypertension.

Conclusion

CMR of the pulmonary vasculature is becoming increasingly one area of investigation in the proximal pulmonary artery. Main pulmonary artery blood flow velocity at peak systole, maximal systolic main pulmonary artery cross-sectional area, and patient height and weight have been used to calculate pulmonary arterial pressure. Main pulmonary artery area distensibility and relative area change, calculated as relative cross-sectional area change, predict mortality in patients with PAH. At present, only limited evidence exists in support of either CT or magnetic resonance angiography for definition of thromboembolic obstructions in chronic thromboembolic pulmonary hypertension.

Suggested Reading

Badesch DB, Champion HC, Sanchez MA. Diagnosis and assessment of pulmonary arterial hypertension. J Am CollCardiol. 54, 2009: S55-S66.009.

Case 6

Patient came with complaints of chest discomfort. Echocardiography was advised which showed aortic dissection involving ascending aorta. CT angiography was advised for further evaluation.

CT Angiography Findings

Type-A dissection with evidence of dissecting flap extending from aortic root involving origin of RCA extending cranially through sinotubular junction and involving aortic arch, origin and proximal portion of left subclavian artery (Figs 12.18, 12.19 and 12.20).

The right brachiocephalic artery, left common carotid arteries were normal in calibre.

Descending thoracic and abdominal aorta showed normal course and calibre with coeliac and mesenteric arteries showing normal course and calibre (Figs 12.21a and b).

Atherosclerotic changes were seen in renal arteries.

Origin of left coronary arteries was normal. There was no evidence of dissection flap extending in the left main artery. Aortic measurements were normal.

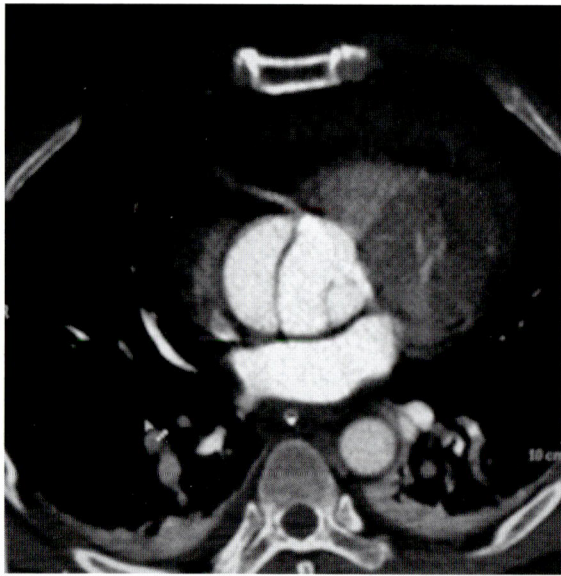

Fig. 12.18: Dissecting flap in aortic root involving origin of RCA and spared left main artery

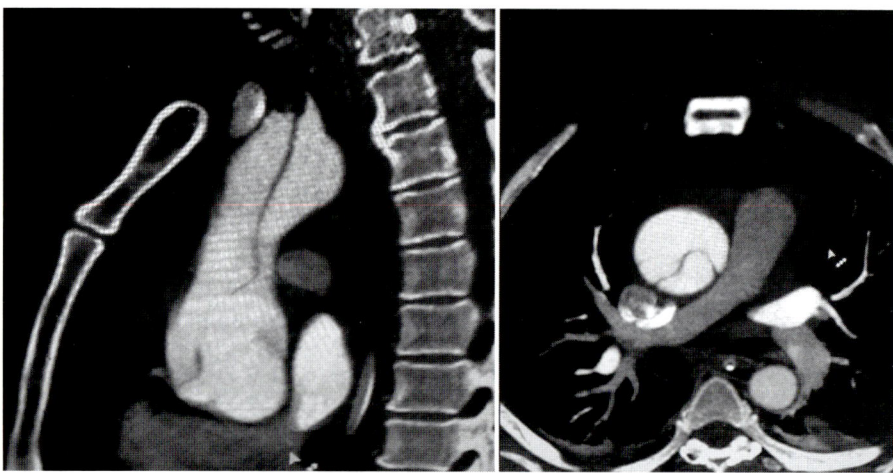

Fig. 12.19: Dissection in ascending aorta

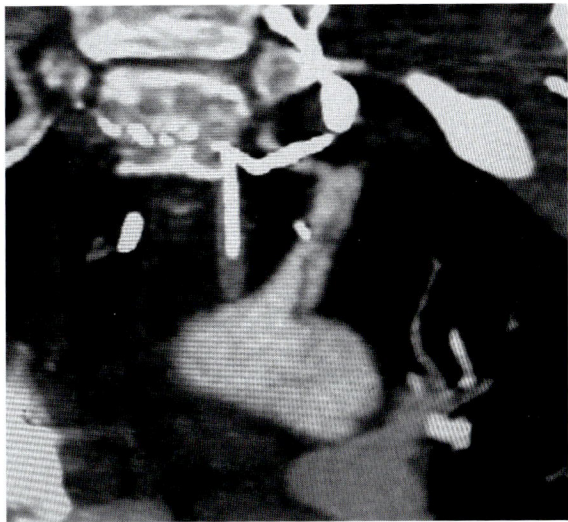

Fig. 12.20: Extension of dissection in left SCA

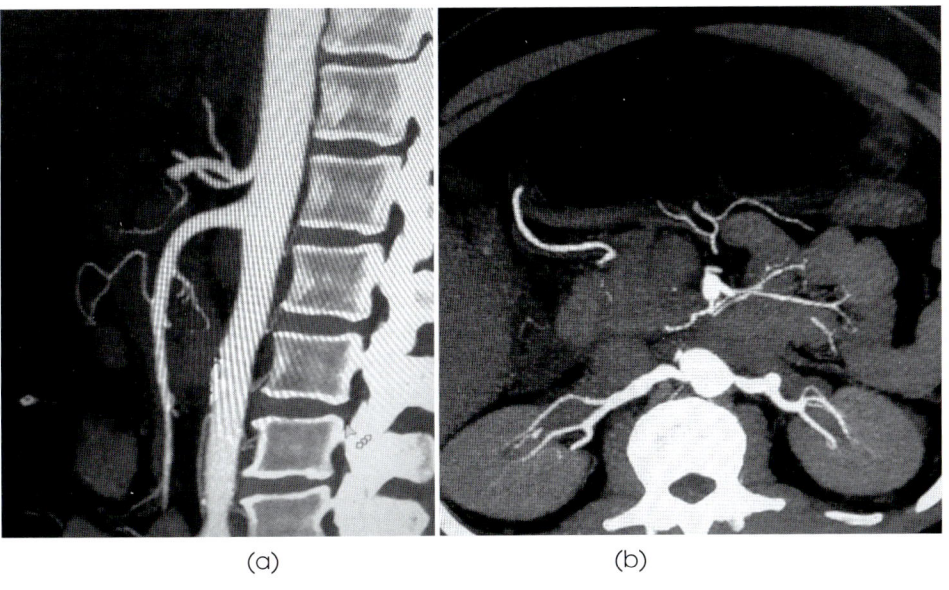

(a)   (b)

Fig. 12.21: Normal mesentric arteries and mild atherosclerotic changes in renal arteries, however, no evdence of dissection noted

CT angiography is the modality of choice in evaluation of aortic dissection. It not only identify the dissection but it also helpful in giving following details.

1. Extent of dissection
2. Thrombus in the false lumen
3. Branch vessel or coronary artery involvement
4. Aortic insufficiency
5. Pericardial effusion with or without tamponade
6. Sites of entry and re-entry

Surgical Intervention

Bentall's procedure was done. Imaging findings were matching with the surgical findings. The dissection was involving origin of RCA and dissecting flap was seen extending in left SCA. Patient was discharged successfully.

Conclusion

Aortic dissection is the most common acute emergency condition of the aorta and often leads to the patient's death. Early diagnosis and treatment are essential for improving the prognosis. It is recommended that the scanning field includes the entire aorta and pelvic vessels to help determine the type and extent of dissection and to help detect complications early enough to improve outcome and aid in treatment planning.

CT imaging of the aorta is fast and widely available, which are important features in making an accurate diagnosis quickly in unstable patients. Multidetector CT allows imaging of the entire aorta with rapid acquisition and data reconstruction to provide prompt and accurate diagnosis and to help identify relevant complications that may have an impact on surgical planning or management.

Suggested Reading

1. Castañer E , Andreu M , Gallardo X , Mata JM , Cabezuelo MA , Pallardó Y. CT in nontraumatic acute thoracic aortic disease: typical and atypical features and complications. Radio Graphics 2003; 23.
2. Mehta RH , Suzuki T , Hagan PG , et al. Predicting death in patients with acute type A aortic dissection. Circulation 2002; 105(2): 200–206.
3. Hagan PG, Nicnaber CA, Isselbacher EM, et al. The International Registry of Acute Aortic Dissection (IRAD): new insights into an old disease. JAMA 2000; 283(7): 897–903.
4. Wheat MW. Acute dissecting aneurysms of the aorta: diagnosis and treatment, 1979. Am Heart J 1980; 99(3): 373–387.

Case 7

Patient was brought by unknown persons. On arrival, he was unconscious, unresponsive, with frank bleeding from nose and mouth.

Past History

Patient was admitted at outside hospital with massive hemoptysis on 03/09/2014, oral and genital ulcers 7 years back. Thrombophlebitis of hand and feet 2 years back, also complaints of blurring of vision for 2 years for which he was prescribed prednisolone which he discontinued 8 months back.

Hospital Course

Immediately patient was resuscitated with standard ACLS protocol had cardiac arrest. CPR done for 5–10 minutes. Patient was revived and put on ventilatory support.

Chest X-ray and CT Pulmonary Angiography Were Done

Findings Were

- On CXR well-defined opacity in bilateral lung fields (Fig. 12.22).
- CT showed B/L pulmonary artery aneurysms which were partially thrombosed (Figs 12.23a and b)
- Bilateral lungs lower lobe infarcts with multiple chest wall collaterals (Figs 12.24, 12.25a and b)
- B/L pulmonary haemorrhages (Figs 12.25a and b)
- CT head was showing normal brain parenchyma (Fig. 12.26).

Radiological Diagnosis

Behçet's syndrome with bilateral partially thrombosed pulmonary artery aneurysms with bilateral lower lobe infarcts with multiple chest wall collaterals.

Behçet's Disease

- Behçet's syndrome is an inflammatory, chronic and systemic disease whose etiology remains unknown, with higher incidence in men aged between 30 and 40 years
- Triad comprises recurrent oral, genital ulcerations and uveitis.
- Pulmonary vasculitis compromises great and medium calibre vessels, most frequently affecting the venous system in 85% of cases, in the form of thrombophlebitis.

Complications

Fibrosing mediastinitis, pulmonary infarction, atelectasis, haemorrhage, diffuse airspace nodules, pneumonia and fibrosis may be observed.

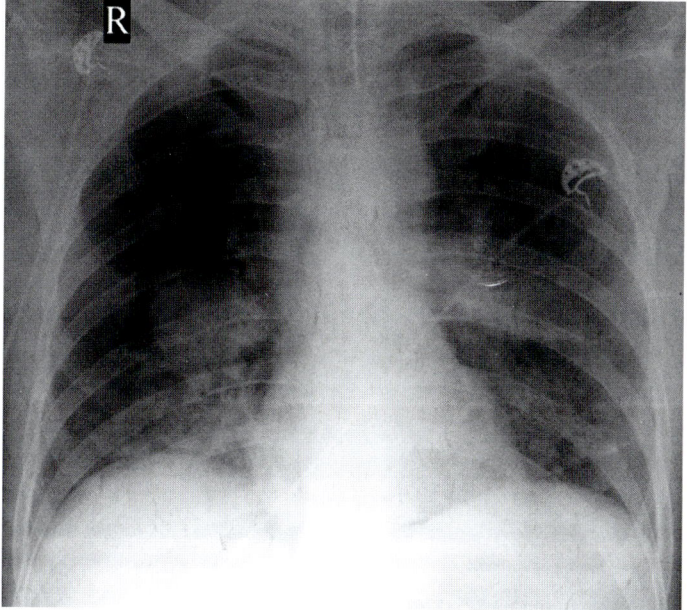

Fig. 12.22: CXR showing well-defined opacities in both lung fields

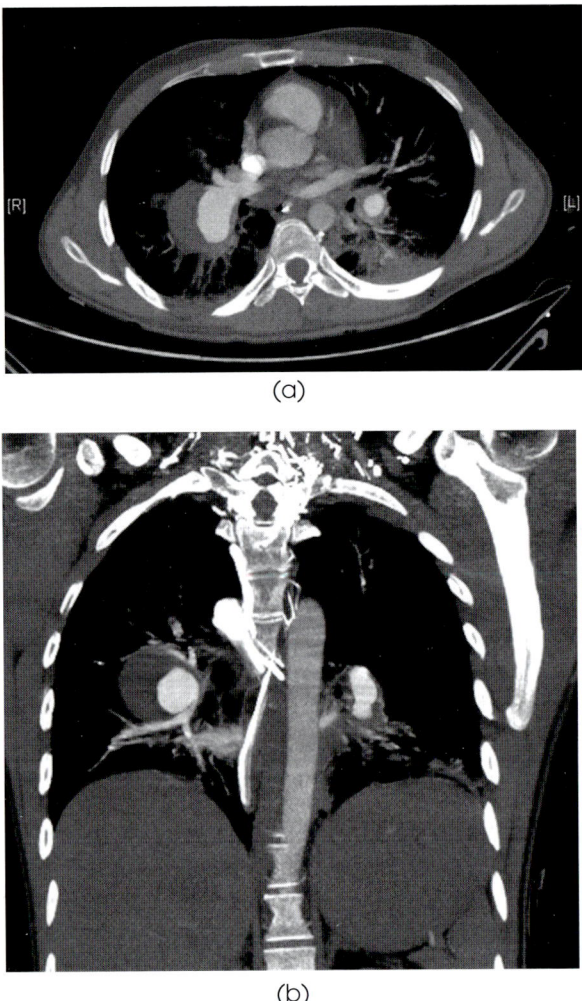

(a)

(b)

Fig. 12.23a and b: CT scan showing partial thrombosed pulmonary artery aneurysms

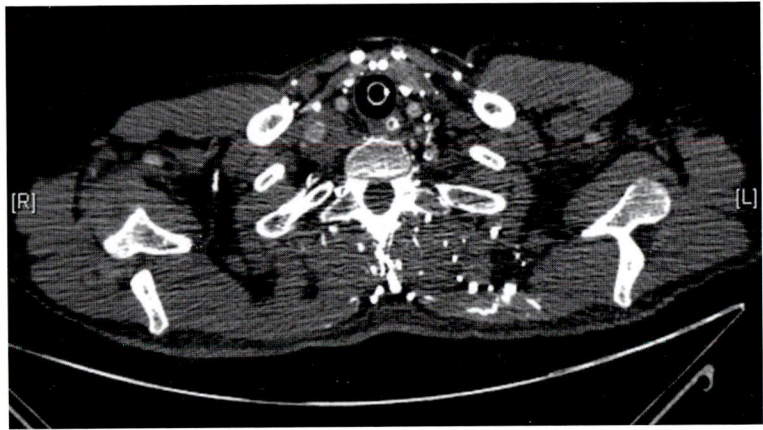

Fig. 12.24: CT scan showing multiple chest wall collaterals

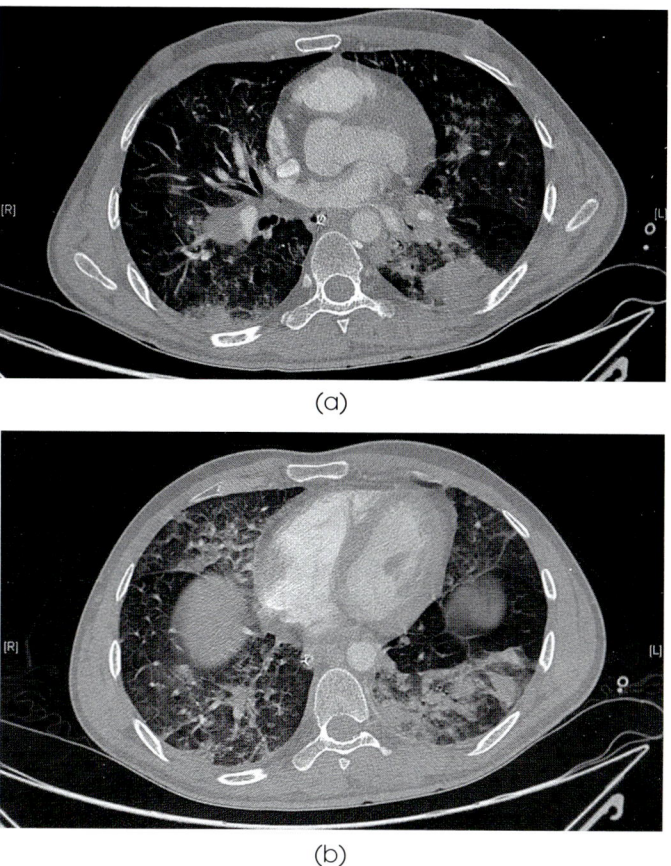

(a)

(b)

Fig. 12.25a and b: Chest CT showing basal infarcts and B/L lung parenchymal haemorrhages

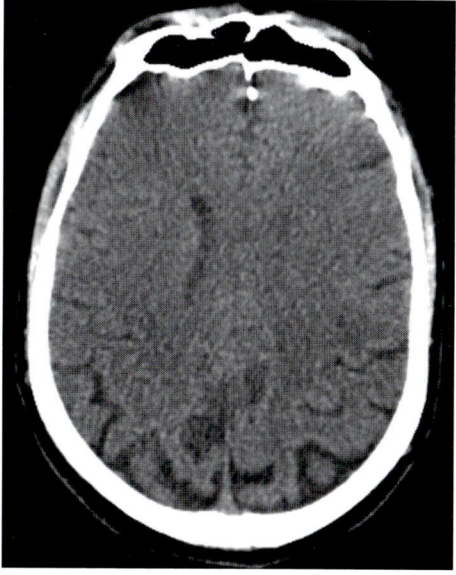

Fig. 12.26: CT head showing normal brain parenchyma

Differential Diagnosis

Hughes-Stovin syndrome

Absence of oral and genital ulcerations is observed.

Surgical Intervention

Pag with angioembolization of right pulmonary artery aneurysms using nestor coils as shown below with serial X-ray up to 5th post-operative day (Figs 12.27a, b, 12.28 and 12.29).

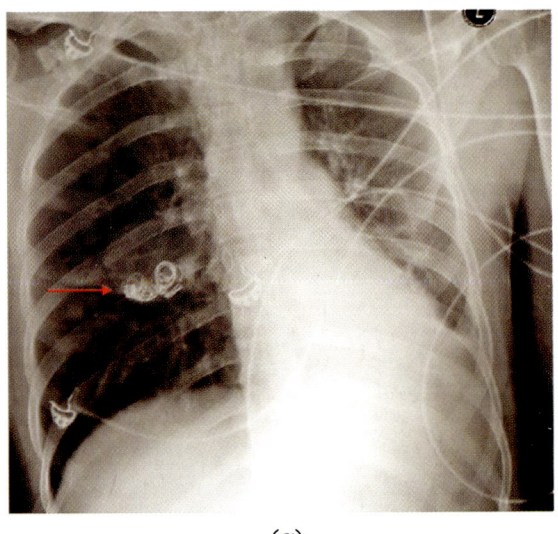

(a)

(b)

Fig. 12.27a and b: Post-operative day 1

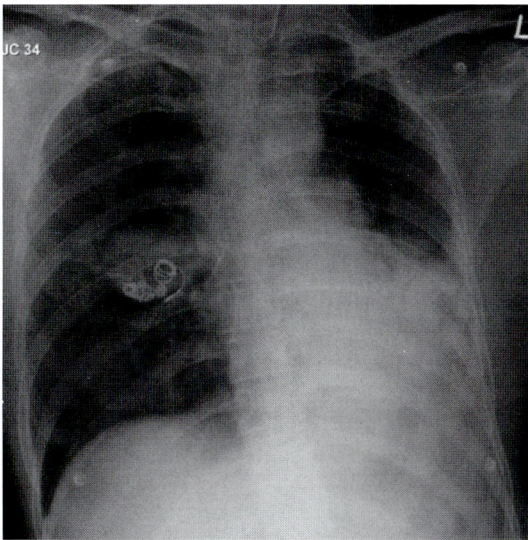

Fig. 12.28: Post-operative day 2

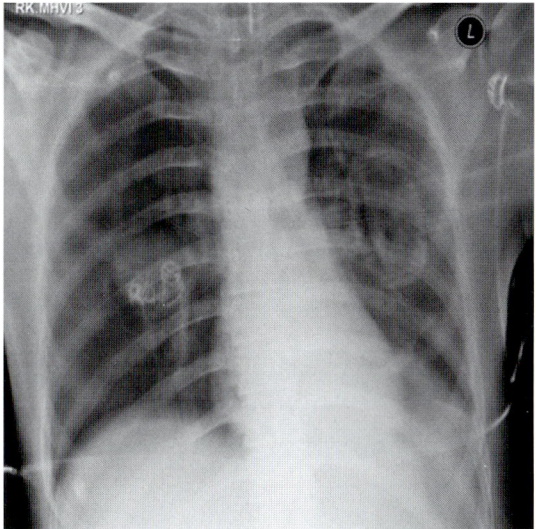

Fig. 12.29: Post-operative day 5

Conclusion

Behçet syndrome is not a common disorder but commonly involves the gastrointestinal tract, with radiologic manifestations similar to those of inflammatory or neoplastic diseases. Familiarity with the radiologic findings helps in making an early diagnosis, as well as in establishing an appropriate treatment strategy.

Suggested Reading

Ha HK, Lee HJ, Yang SK, et al. Intestinal Behçet syndrome: CT features of patients with and patients without complications. Radiology 1998; 209: 449–454.

Case 8

Patient is 52-year-old female with H/o dyslipidemia. Echocardiography was done which showed bicuspid aortic valve and ascending aortic aneurysm. CT aortogram was advised for further evaluation.

CT Aortogram Findings

- Fusiform dilatation of ascending aorta with no evidence of dissection and bicuspid aortic valve (Figs 12.30a, b and 12.31)
- Bilateral coeliac, superior and inferior mesentric arteries were normal
- Origins of coronary arteries were normal (Fig. 12.32a and b)
- Right main renal artery was seen arising at level of SMA and an accessory right renal artery was seen more inferiorly (Fig. 12.33).

Aortic Measurements Were

- Sinus of valsalva—4.2 cm ,
- Sinotubular junction—3.9 cm
- Ascending aorta—5.3 cm, mid-aortic arch—3.3 cm
- Aorta distal to left subclavian artery—2.9 cm
- Descending thoracic aorta—2.3 cm
- Prerenal abdominal aorta—2.2 cm
- Post-renal abdominal aorta—1.7 cm

Radiological Diagnosis

Ascending aortic aneurysm with no evidence of dissection and with associated bicuspid aortic valve.

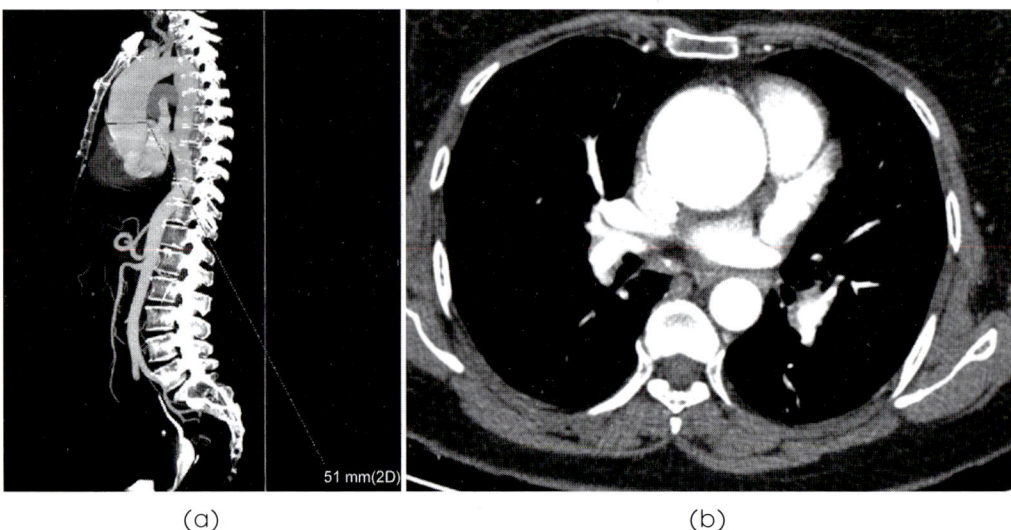

(a)   (b)

Fig. 12.30a and b: CT angiography showing ascending aortic aneurysm

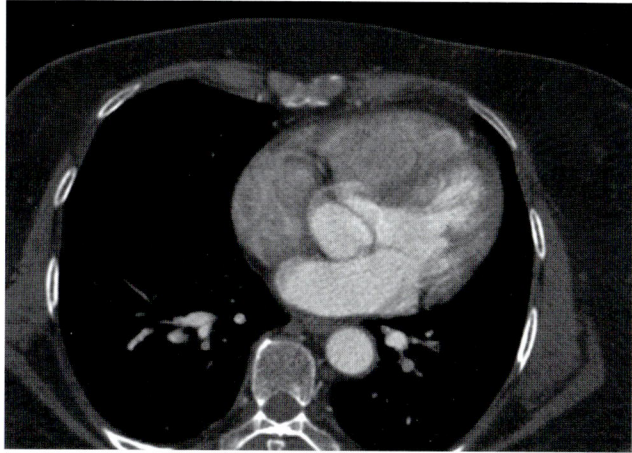

Fig. 12.31: CT showing bicuspid aortic valve

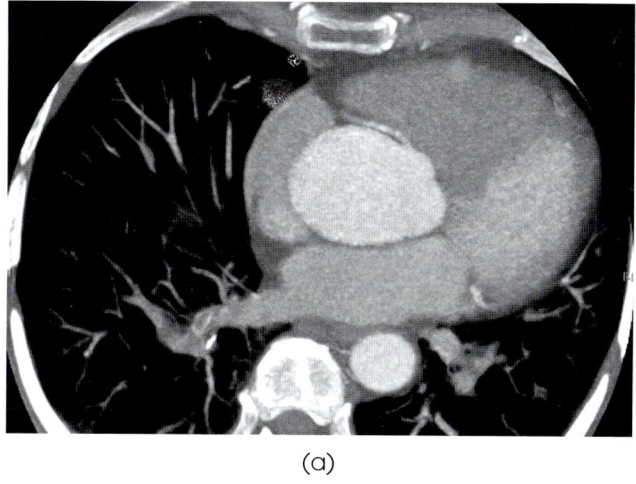

(a)

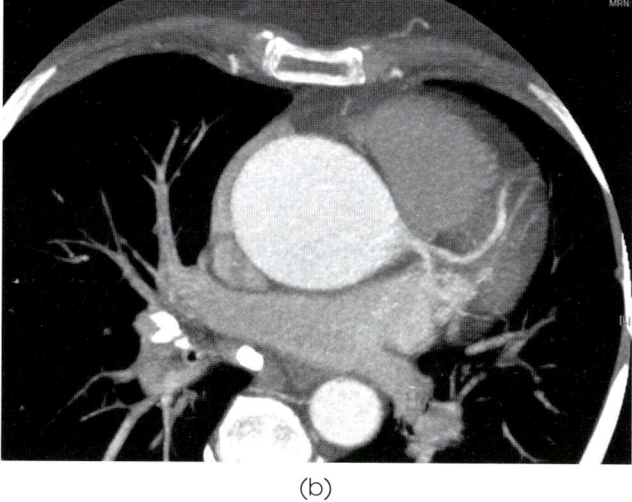

(b)

Fig. 12.32a and b: CT showing normal origin of coronary arteries

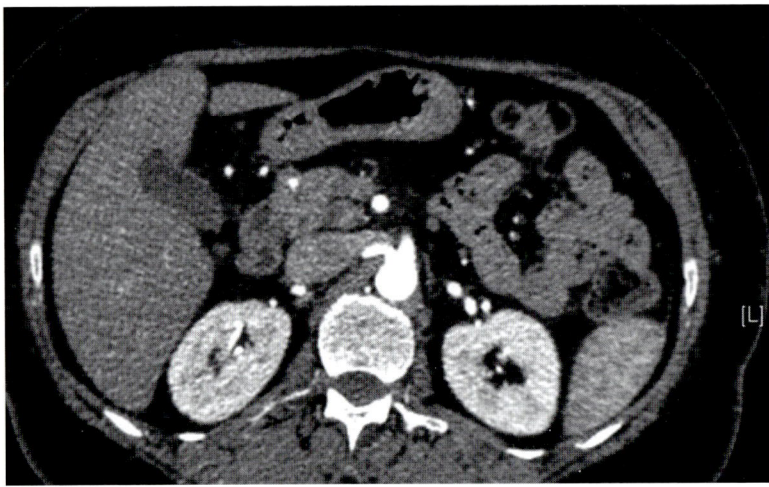

Fig. 12.33: CT showing origin of right renal artery at level of SMA

Surgical Intervention

Bentall's procedure

Conclusion

Multidetector CT angiography is routinely used to evaluate the spectrum of thoracic aortic aneurysms (TAAs). Knowledge of the causes, significance, imaging appearances, and potential complications of both common and uncommon aortic aneurysms is essential for prompt and accurate diagnosis.

Suggested Reading

1. Rajagopalan S, Sanz J, Ribeiro VG, Dellegrottaglie S. CT angiography of the thoracic aorta with protocols. In: Mukherjee D, Rajagopalan S, eds. CT and MR angiography of the peripheral circulation: practical approach with clinical protocols. London, England: Informa Healthcare, 2007; 91–110.
2. Green CE, Klein JF. Multidetector row CT angiography of the thoracic aorta. In: Boiselle PM, White CS, Eds. New techniques in cardiothoracic imaging. New York, NY: Informa Healthcare, 2007; 105–126.

Case 10

Patient has history of pain in upper back since 1 year with weight loss. No histoy of fever. ATT started on 1/5/2014. Underwent treatment for 5 months and his symptoms had improved. 3 months later, the back pain returned. The patient had paraparesis with bladder and bowel involvement, so MRI spine was advised. It was found that there was paravertebral collection. A paraspinal swelling was noted at D8-10 level. Screening for CT guided biopsy revealed that the collection was encasing the aorta. Then CT angiography was advised.

Past History

- Chest TB 18 years back (ATT taken).
- K/C/O DM on treatment.

- Post PTCA (stent to LAD in May 13)
- Hypertension.

CT Angiography findings

1. Saccular lobulated aneurysms in supra renal aorta (Figs 12.34a, b, c and 12.35)
2. Soft tissue was seen surrounding the aneurysm with underlying bony erosion and sclerosis of vertebra (Figs 12.36a, b, 12.37a and b).

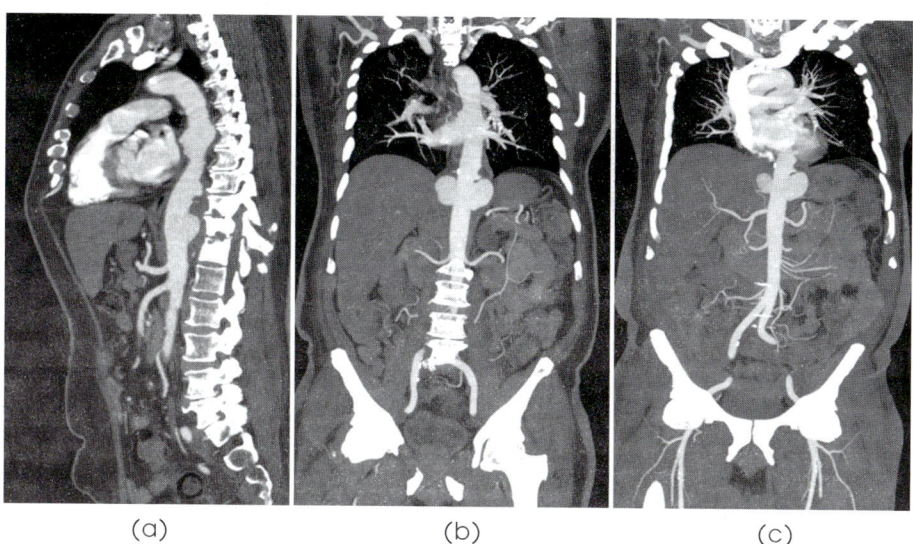

(a)   (b)   (c)

Fig. 12.34a to c: CT angiography reveals saccular lobulated aneurysms in suprarenal aorta

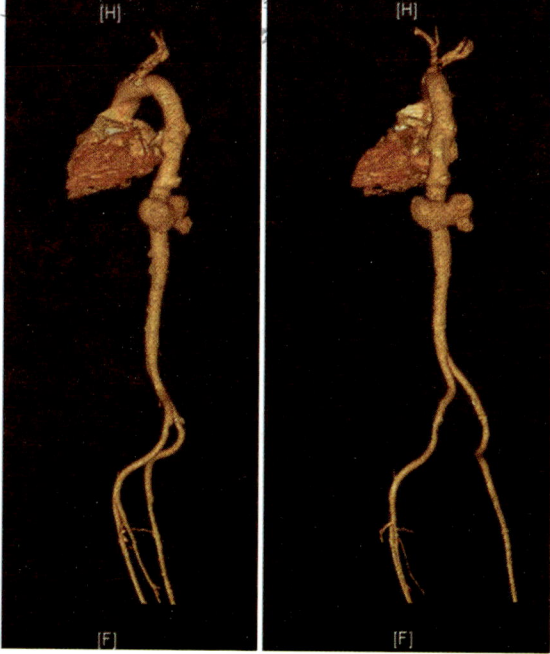

Fig. 12.35: CT volume rendering image

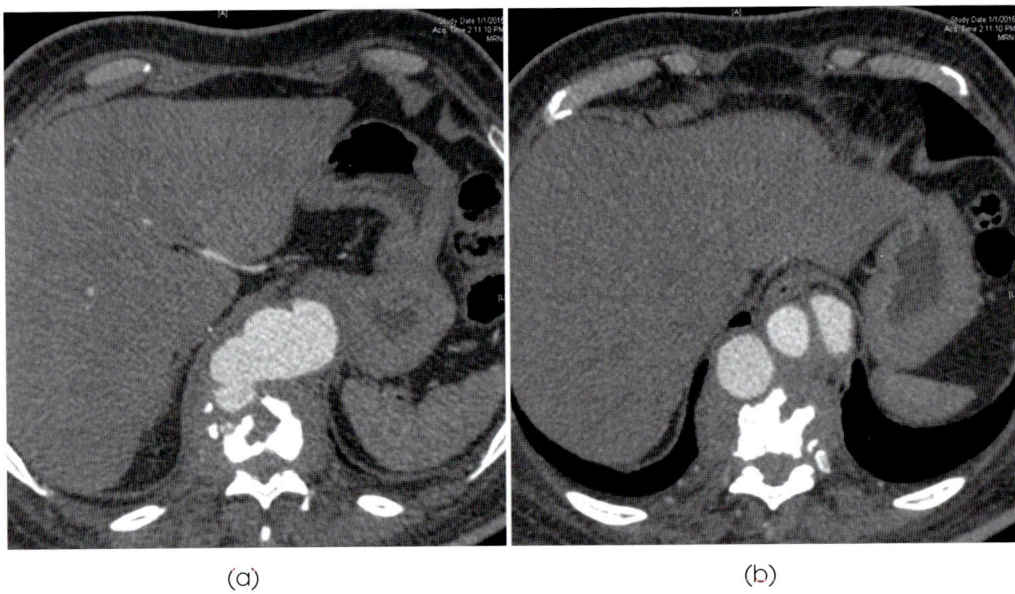

(a)   (b)

Fig. 12.36a and b: CT showing soft tissue surrounding the aneurysm and the underlying erosion of vertebra

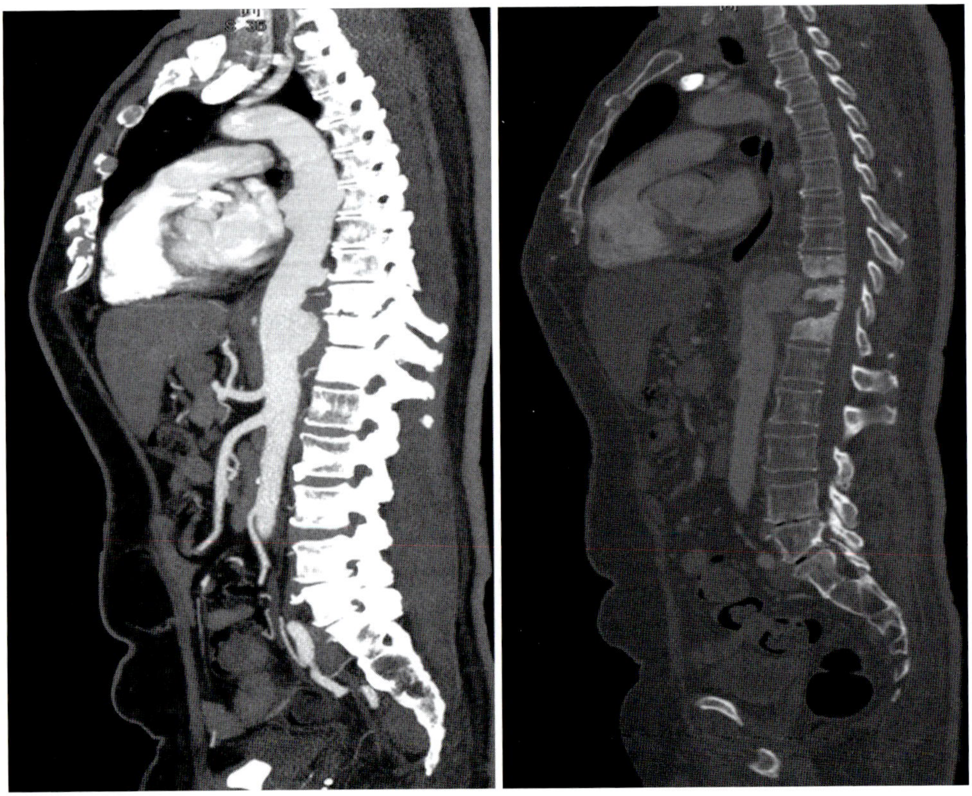

(a)   (b)

Fig. 12.37a and b: CT saggital reformat showing bony erosion and sclerosis of vertebrae

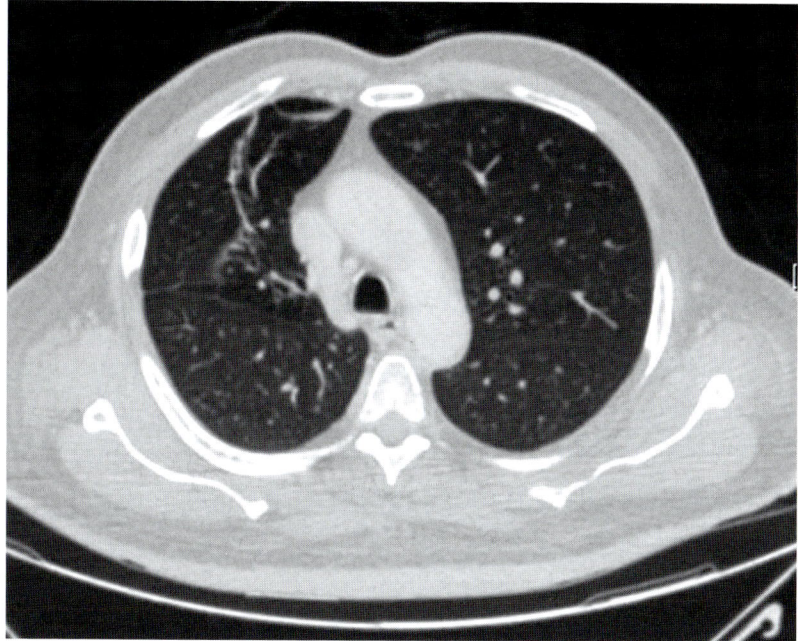

Fig. 12.38: CT showing fibrosis in right upper lobe

Radiological Diagnosis

Mycotoc aneurysm

Surgical Intervention

- Thoracic endovascular aortic aneurysm repair (5/1/15).
- Orthopaedic: Posterior decompression, *trans*-pedicular decancellation, and left pedicle excision at D10, with D9-10 interbody fusion on 19 Jan, 2015).

Conclusion

CT angiography is very helpful in diagnosis, disease progression and to describe the complication of mycotic aneurysm.

Suggested Reading

1. Stavropoulos SW, Charagundla SR. Imaging techniques for detection and management of endoleaks after endovascular aortic aneurysm repair. Radiology 2007; 243(3): 641–655.
2. Fillinger MF. Postoperative imaging after endovascular AAA repair. Semin Vasc Surg 1999; 12(4): 327–338. Medline.

Case 11

PT is known case of DM, smoker, alcoholic, CAD, had sudden onset breathlessness came to the hospital advised CAG and echocardiography CAG showed—tripple vs disease and echo showed moderate MR, LVEF—30–35%, akinetic basal mid-inferior, posterior, lateral wall.

Patient was Advised for CT Neck Angiography

CT neck angiography findings:

1. Right ICA origin shows stenosis in Figs 12.39, 12.41a and b.
2. Left subclavian artery origin stenosis (Figs 12.40a, b and 12.41a).

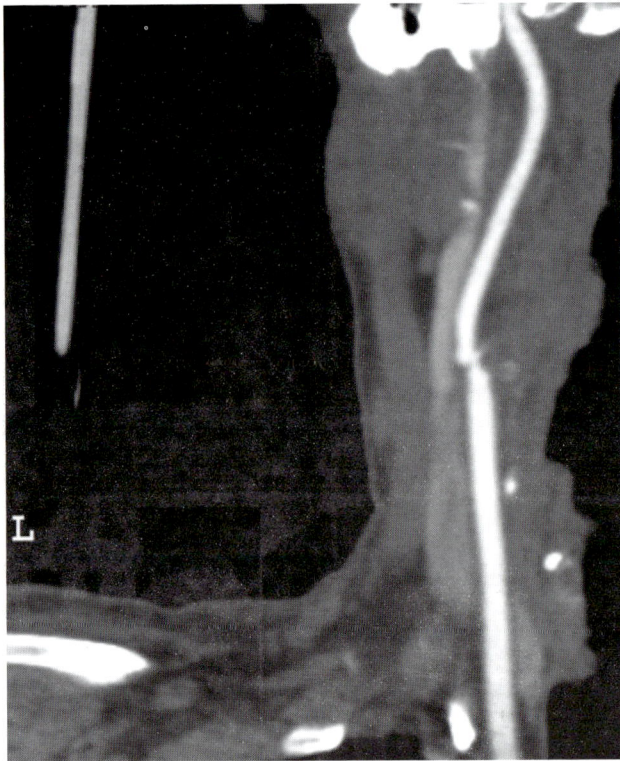

Fig. 12.39: CT angiography showing right ICA origin stenosis

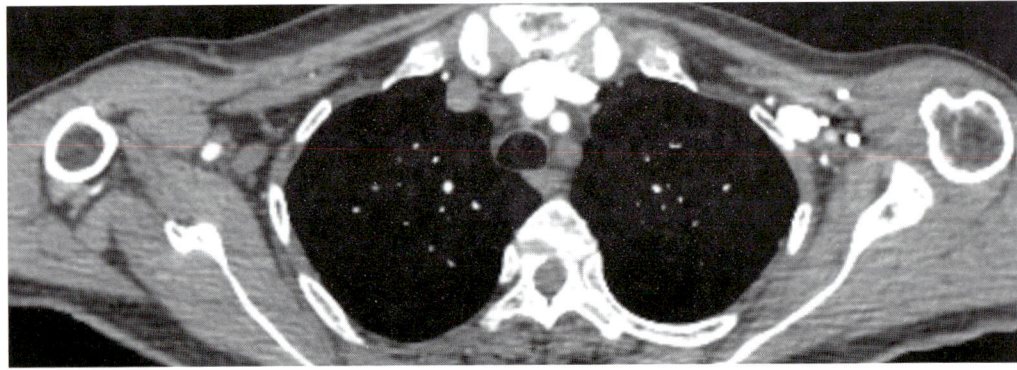

Fig. 12.40a: CT angiography showing left subclavian artery origin stenosis

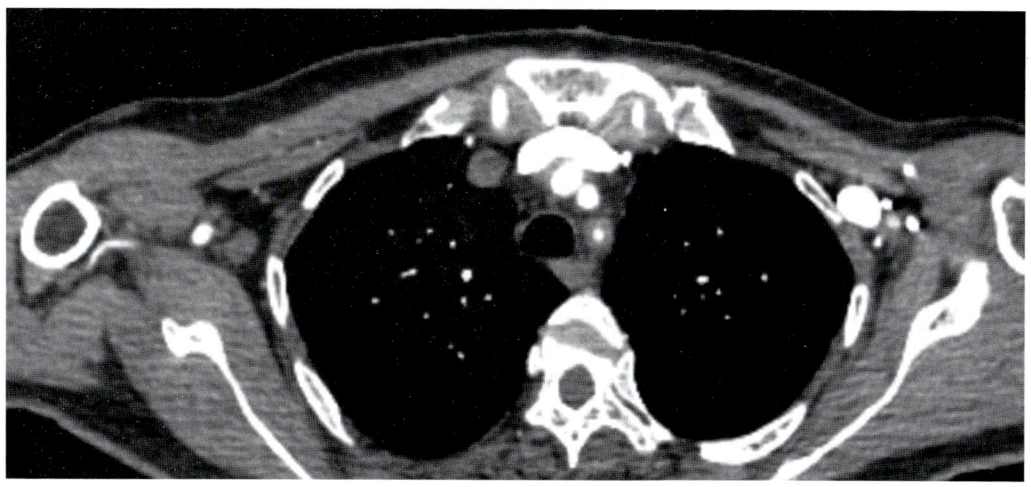

Fig. 12.40b: Left subclavian artery stenosis

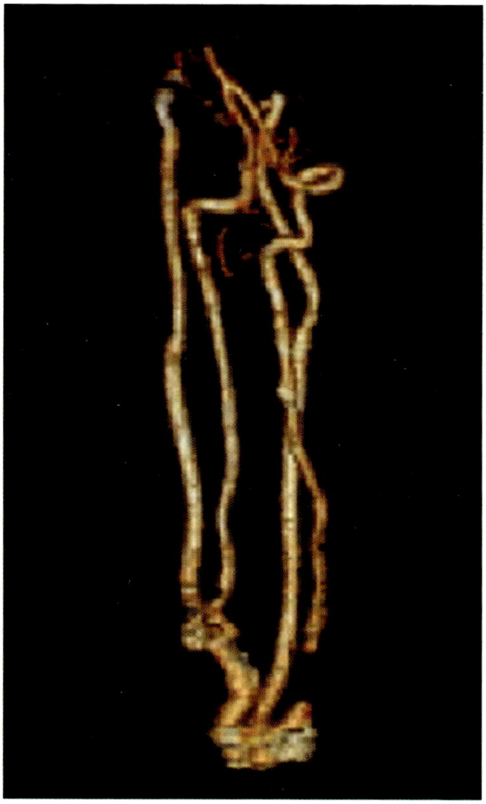

(a) 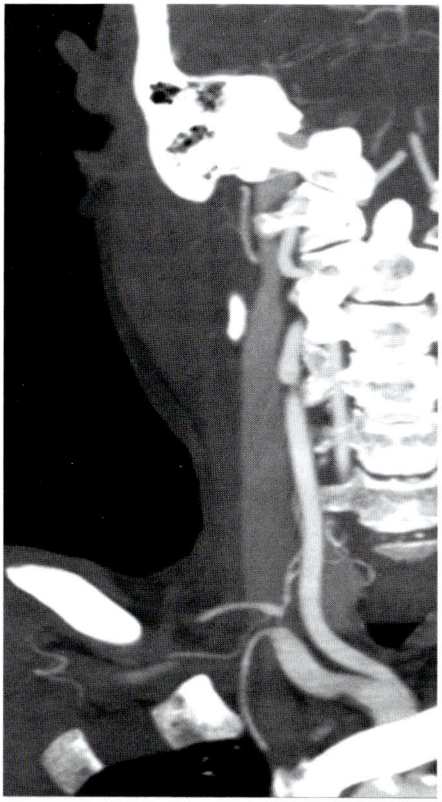   (b)

Fig. 12.41a and b: CT volume rendering and MPR image

Surgical Intervention

CABG done with RSVG to LAD, RSVG to diagonal, RSVG to OM and RSVG to RCA. LIMA was not harvested due to left subclavian stenosis.

Conclusion

MR angiography is used increasingly in imaging of carotid artery stenosis. In the past decade, findings of several studies have been published in which MR angiography was compared with DSA. The most commonly applied technique in these studies was TOF MR angiography. In the past few years, implementation of enhanced MR angiography has been repeatedly suggested in the diagnosis of carotid artery stenosis. Enhanced MR angiography of the carotid arteries is a recent development that could minimize signal loss and motion artifacts.

Suggested Reading

1. Hankey GJ, Warlow CP, Sellar RJ. Cerebral angiographic risk in mild cerebrovascular disease. Stroke 1990; 21: 209–222.
2. Bendszus M, Koltzenburg M, Burger R, Warmuth-Metz M, Hofmann E, Solymosi L. Silent embolism in diagnostic cerebral angiography and neurointerventional procedures: a prospective study. Lancet 1999; 354: 1594–1597.
3. De Marco JK, Nesbit GM, Wesbey GE, Richardson D. Prospective evaluation of extracranial carotid stenosis: MR angiography with maximum-intensity projections and multiplanar reformation compared with conventional angiography. AJR Am J Roentgenol 1994; 163: 1205–1212.

Case 12

Patient is 60-year-old male presented with history of left shoulder pain. Patient's status is post-operative dissecting thoracic aorta (2007, 2010) with carotid-carotid grafting and bilateral renal artery stenting (Figs 12.42a, b, 12.43a and b).

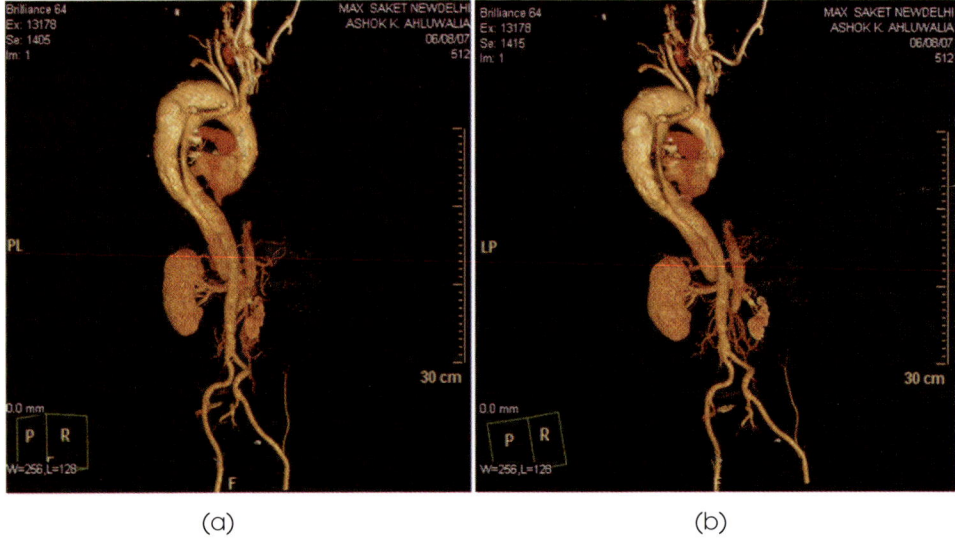

(a)   (b)

Figs 12.42a and b: MRI previous VR images aortic dissection 2007

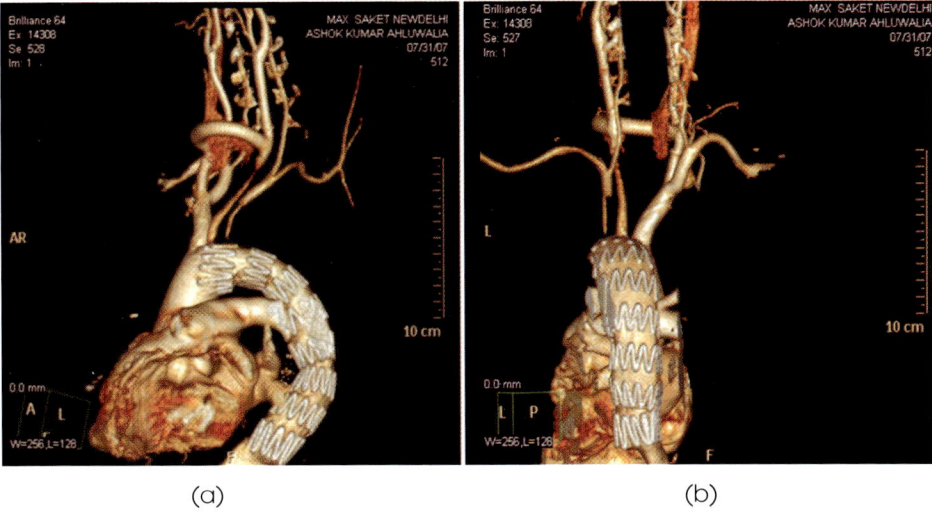

(a)   (b)

Fig. 12.43a and b: 2007 Carotid-carotid grafting with aortic stenting

Now patient came with complaints of chest pain and admitted for further evaluation. CXR and CT angiography were done.

Findings

1. Blocked carotid graft and patent aortic graft noted (Figs 12.44 and 12.45).
2. Large perigraft thrombus displacing pulmonary artery and vein compared with previous CT angiography of 2010 there was no change in size of perigraft thrombus and no obvious leak of contrast noted (Figs 12.46a, b, 12.47a and b).
3. Dissection was extended in descending abdominal aorta up to renal level (Figs 12.48 and 12.49).
4. Coeliac artery taking origin from false and SMA from true lumen (Figs 12.50 and 12.51).
5. Stented renal arteries which looks patent (Fig. 12.52).

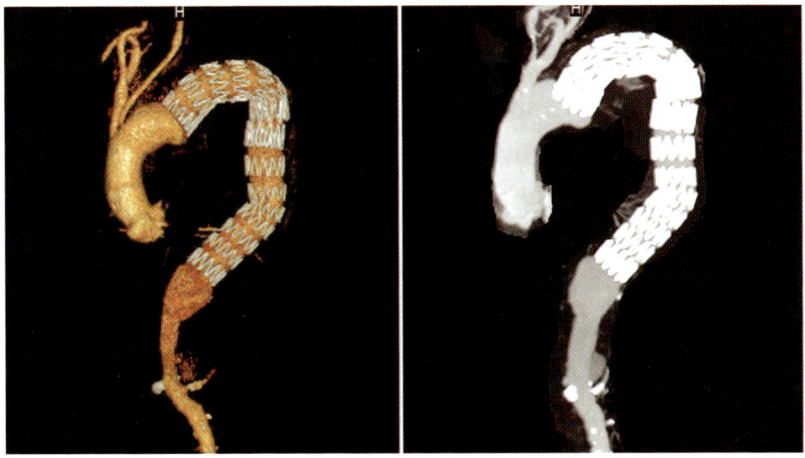

Fig. 12.44: CT volume rendering and reformatted image showing aortic graft

6. Infrarenal aorta showed atherosclerotic plaques, however, there was no intraluminal stenosis (Fig. 12.53a).
7. Iliac arteries were normal calibre (Fig. 12.53b).

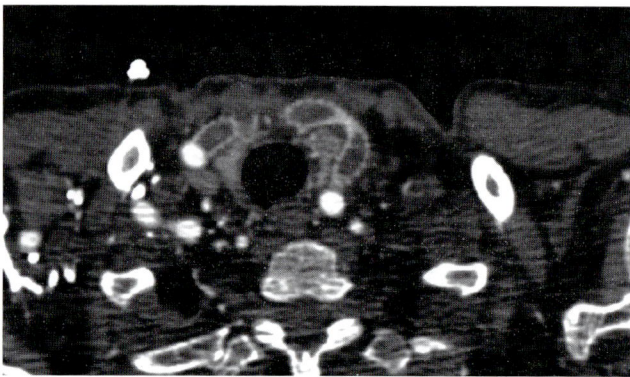

Fig. 12.45: CT showing blocked carotid-carotid graft

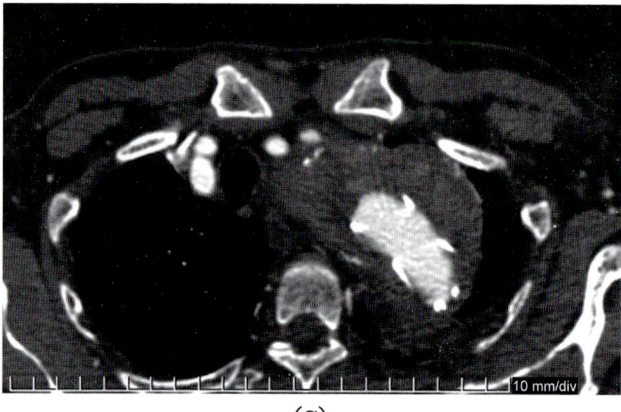

(a)

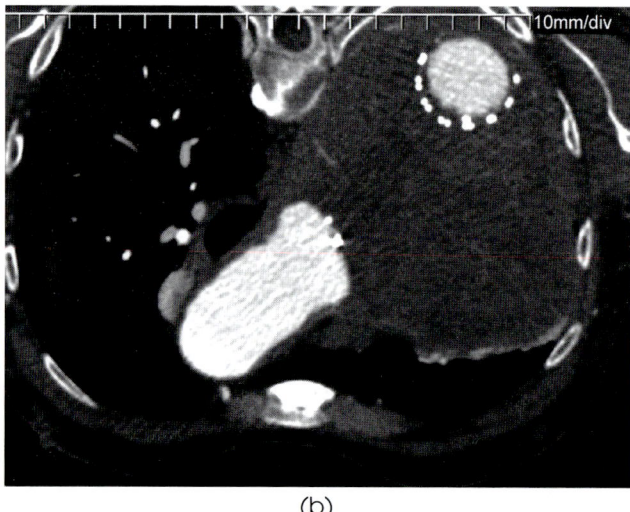

(b)

Fig. 12.46a and b: CT showing large perigraft thrombus

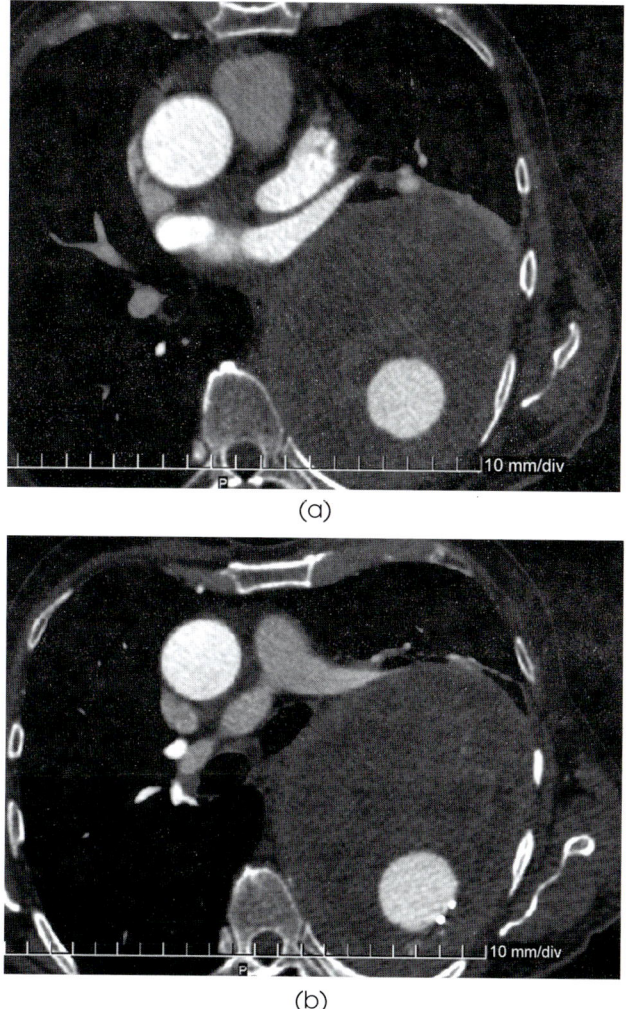

(a)

(b)

Fig. 12.47a and b: CT showing perigraft thrombus displacing the pulmonary artery and vein

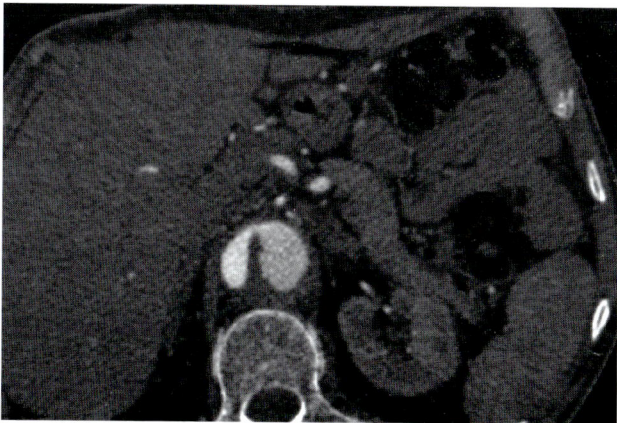

Fig. 12.48: Infradiaphragmatic and suprarenal dissection

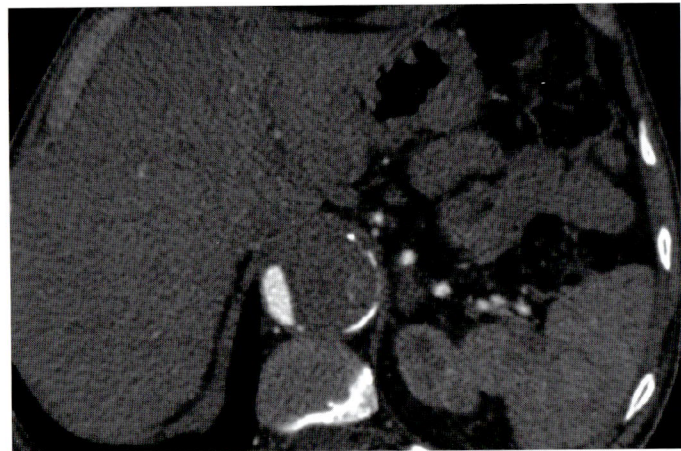

Fig. 12.49: Dissection showing true and false lumen

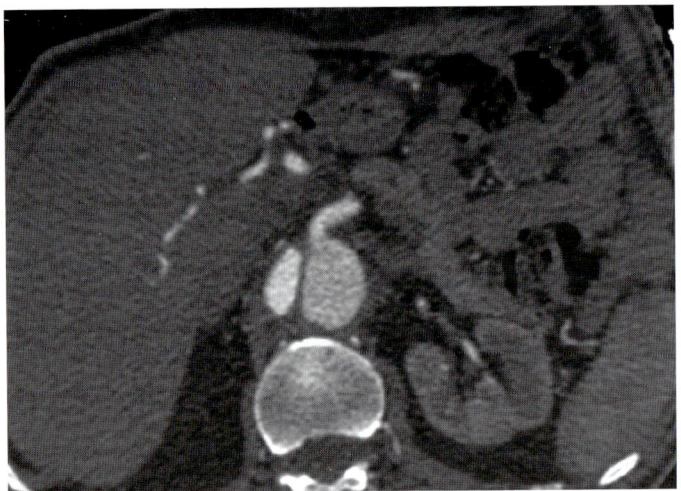

Fig. 12.50: CT showing origin of coeliac artery from false lumen

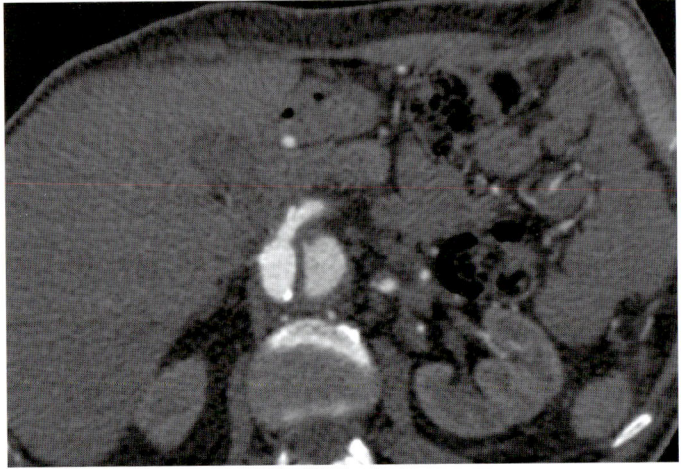

Fig. 12.51: CT showing origin of SMA from true lumen

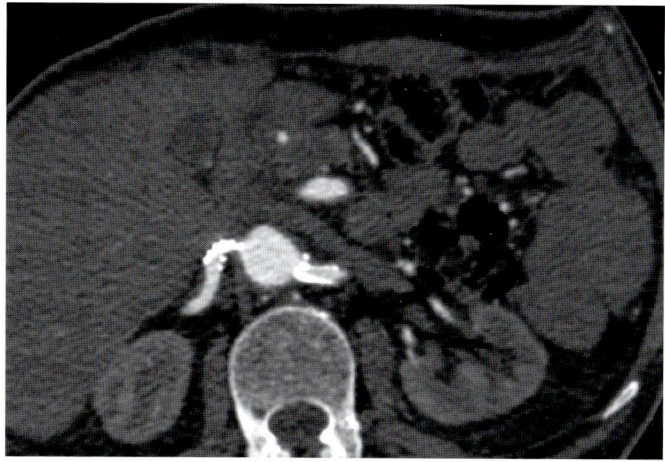

Fig. 12.52: Stented renal arteries

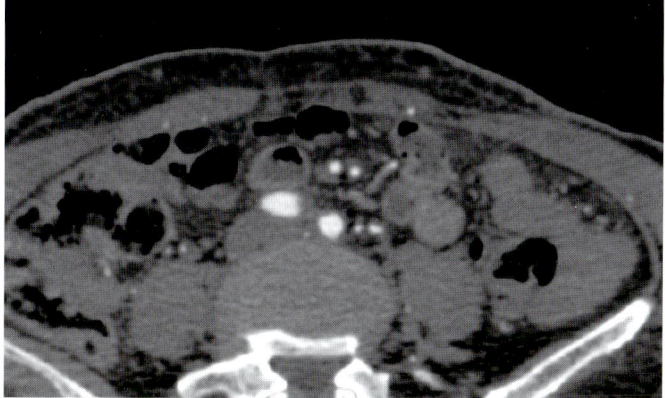

Fig. 12.53a: CT at aortic bifurcation

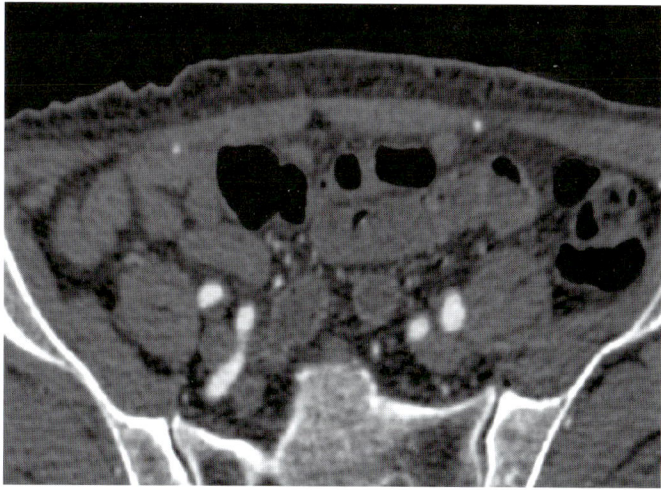

Fig. 12.53b: CT showing normal iliac arteries

Radiological Diagnosis

Blocked carotid graft, large perigraft thrombus displacing pulmonary artery and vein. As compared with previous CT angiography of 2010 there was no change in size of perigraft thrombus and no obvious leak of contrast noted.

Patent stented renal arteries.

Conclusion

CT findings suggestive of complications occurred at any time up to 18 years after surgery. The key to accurate diagnosis and evaluation of a suspected endoleak is multiphase imaging (typically three-phase: non-contrast, arterial phase and delayed phase). Non-contrast is necessary to establish a baseline of density within the (hopefully) thrombosed sacs, presence of calcification which can mimic contrast aging.

- Leakage was defined as the persistence of blood flow outside the lumen of the endoluminal graft but within the aneurysm sac.
- Presence, extent, and origin of leaks.
- Leakage that occurred during the perioperative period was defined as primary endovascular leakage.
- Endovascular leakage that occurred as a late event after successful stent-graft implantation was defined as secondary endovascular leakage.
- Leakage associated with retrograde flow from collateral arterial branches (non-graft-related leakage).

Type of Leak

- *Type I:* Leak at graft attachment site
  - I: Proximal
  - I: Distal
- *Type II:* Aneurysm sac filling via branch vessel (most common)
- *Type III:* Leak through defect in graft
- *Type IV:* Leak through graft fabric as a result of graft porosity.

Suggested Reading

1. Rosen RJ, Green RM. Endoleak management following endovascular aneurysm repair. J Vasc Interv Radiol. 2008; 19 (6 Suppl): S37-43. doi:10.1016/j.jvir.2008.01.017 - Pubmed citation
2. Kaufman JA, Lee MJ. Vascular and interventional radiology, the requisites. Mosby Inc. (2004) ISBN: 0815143699. Read it at Google Books. Find it at Amazon.

Case 13

Patient underwent intravenous cannulation and subsequently after it patient had swelling in right arm which was increasing in size.

CT Angiography Done

Findings were

1. Large hematoma in right arm region (Fig. 12.54).
2. Extravasation of contrast from brachial artery (BCA) at elbow level (Figs 12.55a, b, 12.56a and b).

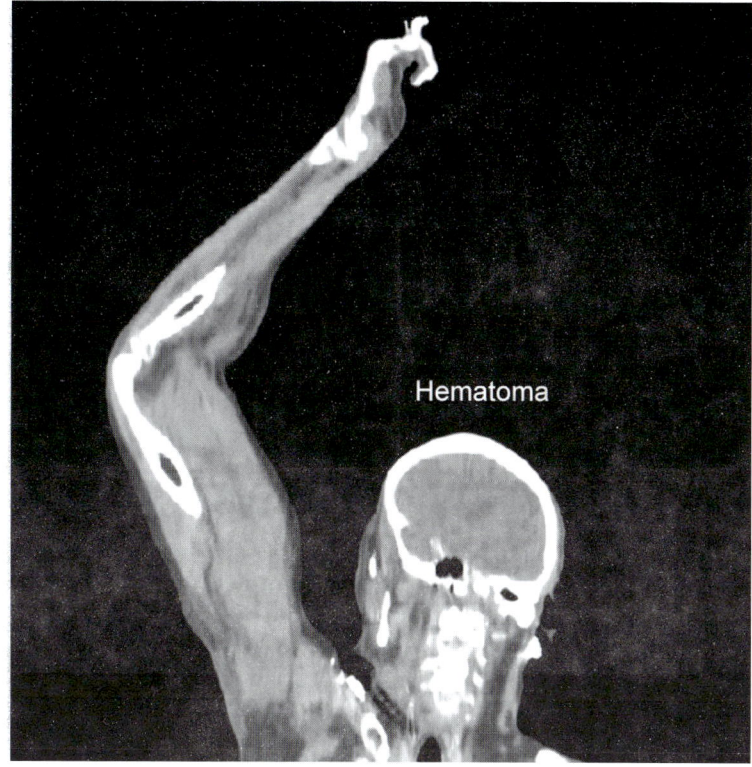

Fig. 12.54: CT showing hematoma in arm region

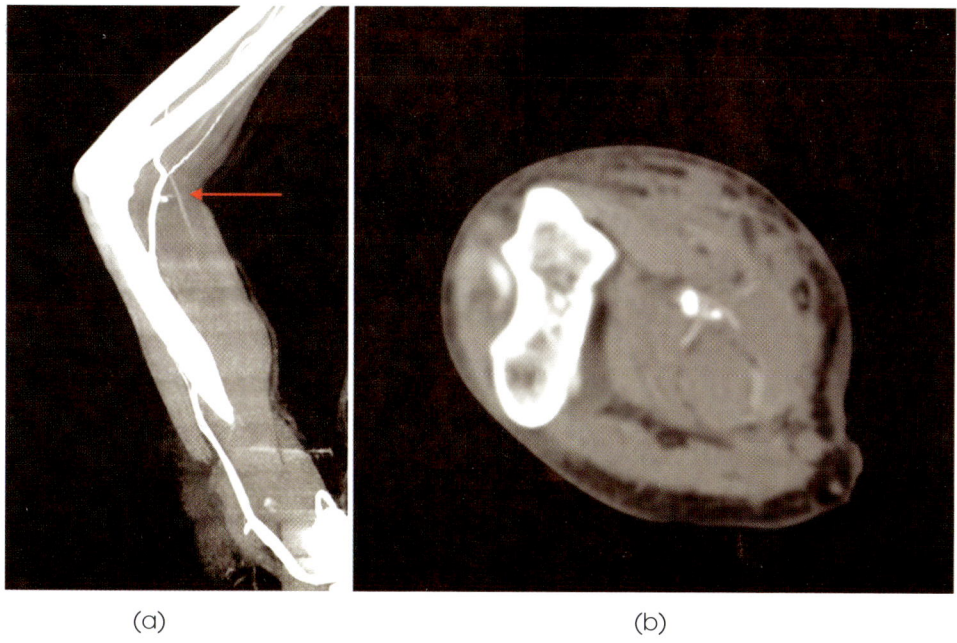

(a)   (b)

Fig. 12.55: CT showing small leaking of intravenous contrast from BCA

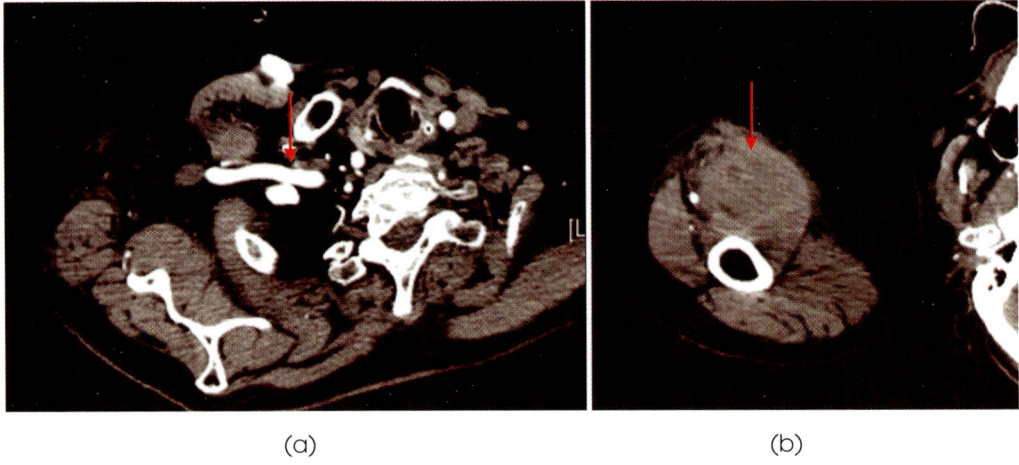

(a)   (b)

Fig. 12.56: CT showing RT axillary artery and RT BCA

Radiological Diagnosis

Brachial artery active bleed.

Surgical Intervention

Procedure: Exploration with vascular repair (brachial artery injury vascular repair) under GA on 18/05/15

Conclusion

CT angiography can be used to diagnose arterial injuries to the extremities rapidly and reliably in the setting of trauma, reducing the delay before repair of the injury. CT angiography is performed with multidetector CT scanners and rapid injection of intravenous contrast material to opacify the arteries. Signs of arterial injury include abrupt narrowing of an artery, loss of opacification of an arterial segment, pseudoaneurysm formation, active extravasation of contrast material, and arteriovenous fistula formation. CT angiograms may be rendered nondiagnostic by motion artifact, lack of opacification of the arteries, and metallic streak artifact. The resolution possible with four-detector row CT scanners may not be high enough to image intimal flap injuries and injuries to arteries distal to the elbow and ankle. Studies of CT angiography used to detect traumatic arterial injuries to the extremities have shown its sensitivity to be 90–95.1% and its specificity to be 98.7–100%. Radiologists and clinicians treating patients with trauma should consider using CT angiography to diagnose arterial injuries in the extremities.

Suggested Reading

1. Menzoian JO, Doyle JE, Cantelmo NL, LoGerfo FW, Hirsch E. A comprehensive approach to extremity vascular trauma. Arch Surg 1985; 120: 801–805.
2. Anderson RJ, Hobson RW 2nd, Padberg FT Jr, et al. Penetrating extremity trauma: identification of patients at high-risk requiring arteriography. J Vasc Surg1990; 11: 544–548.

Case 14

Shortness of breath on mild exertion. Patient is known case of HTN, Takayasu's arteritis, left renal artery stenting 2009. Nonfunctioning left kidney, severe AR, non diabetic, no complaints of chest pain, normal CAG. Echo done showed global hypokinesia, EF—20%, mild AS/severe AR and severe MR.

CT Aortogram Done

Findings were

Irregular saccular aneursmal dilatation of infradiaphragmatic suprarenal aorta with partially thrombosed lumen.

1. Small focal outpouching of supradiaphragmatic aorta (Figs 12.57a and b), Fig. 12.60.
2. Dilated left ventricle (Fig. 12.59)

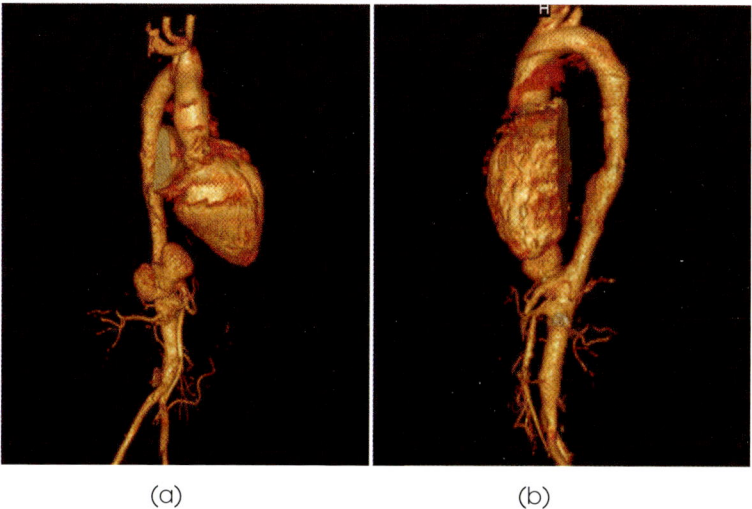

(a)   (b)

Fig. 12.57: CT showing infradiaphragmatic lobulated saccular aneurysm

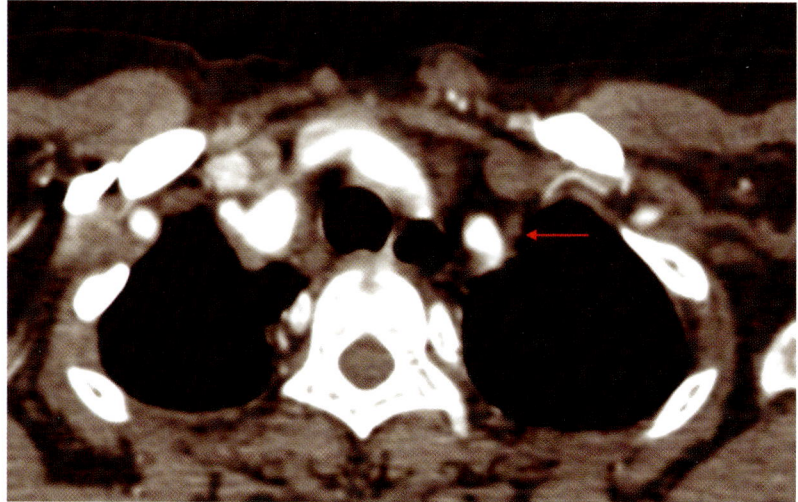

Fig. 12.58: CT angiograpy showing complete block of left subclavian artery

3. Total occluded left subclavian artery (Fig. 12.58).
4. Diffuse mild intimal wall thickening in rest of aorta and iliac arteries and origin of coeliac artery from aneurysmal aortic (Fig. 12.61).
5. Severe stenosis of origin of RT renal artery and stented LT renal artery (Figs 12.62a and b).

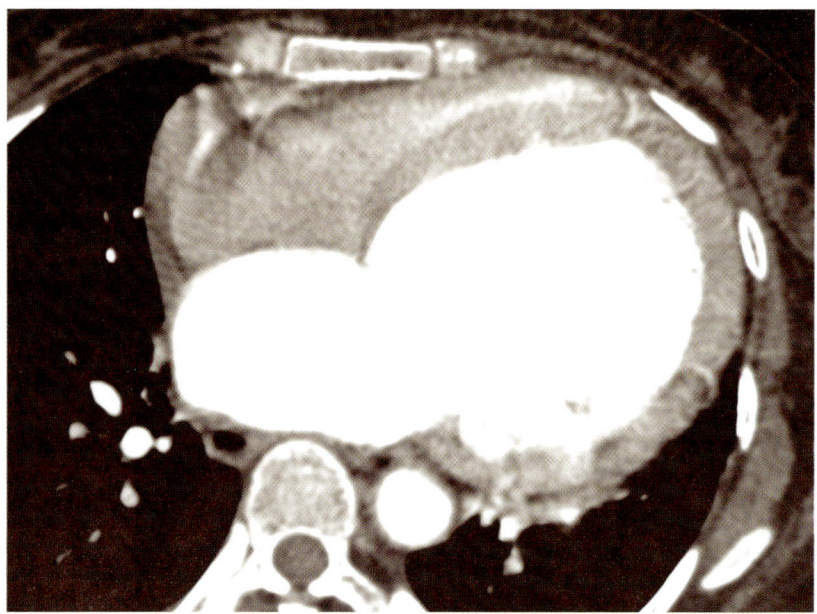

Fig. 12.59: CT showing dilated left ventricle

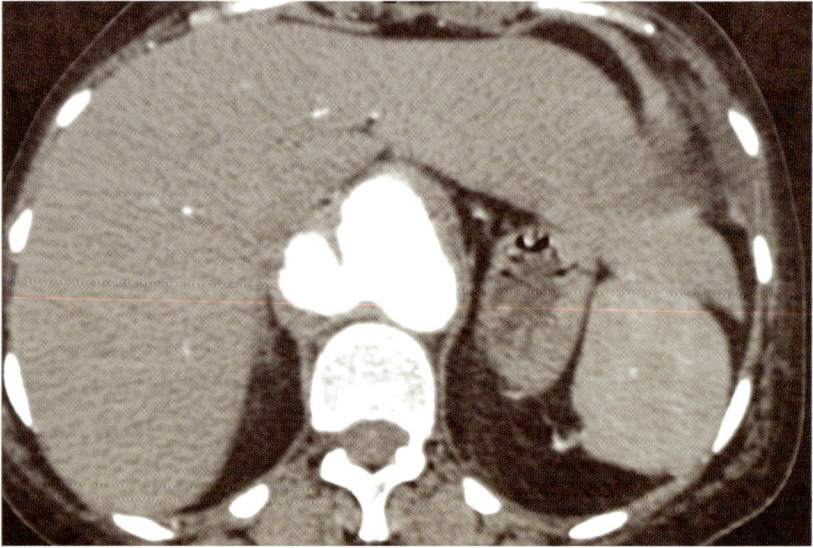

Fig. 12.60: CT angiography showing irregular saccular aneursmal dilatation of infradiaphragmatic supra renal aorta with partially thrombosed lumen

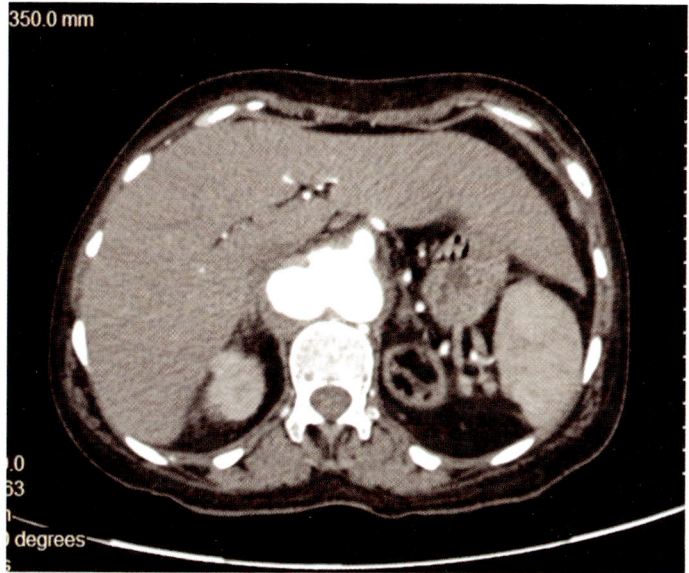

Fig. 12.61: CT showing origin of coeliac artery from aneurysmal aorta

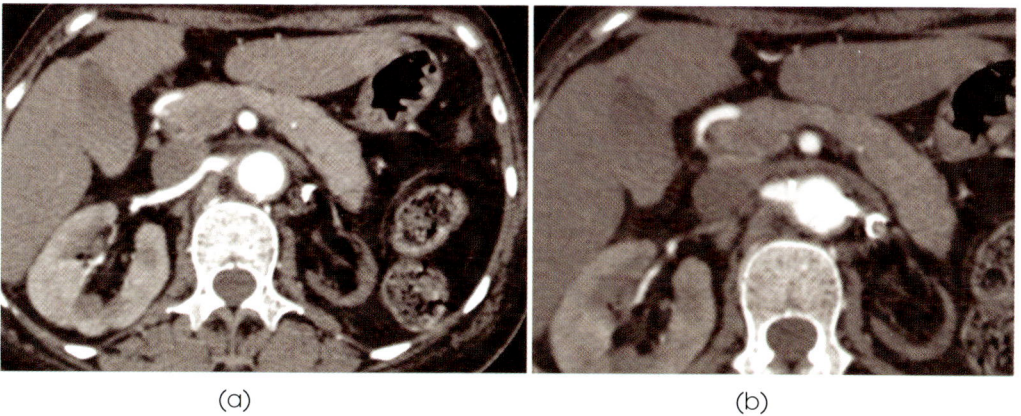

(a)   (b)

Fig. 12.62a and b: CT showing severe stenosis of origin of b/l renal arteries and stented left renal artery

Radiological Diagnosis

Takayasu Arteritis

Conclusion

CT Angiography is very helpful in diagnosis, disease progression and to describe the complications of Takayasu arteritis.

CT Findings Include

- Wall thickening: Acute active phase
- Wall enhancement: Acute active phase
- Aortic valve disease: Stenosis, regurgitation

- Occlusion of major aortic branches
- Aneurysmal dilatation of the aorta or its branches
- Pseudoaneurysm formation
- Diffuse narrowing distally (i.e. descending and abdominal aorta): In late phase

    The pulmonary arteries are also commonly involved, with the most common appearance being peripheral pruning.

Takayasu Arteritis

Takayasu arteritis (TA) (also known as idiopathic medial aortopathy or pulseless disease) is a granulomatous large vessel vasculitis that predominantly affects the aorta and its major branches. It may also affect the pulmonary arteries. The exact cause is not well known but the pathology is thought be similar to giant cell arteritis.

Pathology

There is segmental and patchy granulomatous inflammation of the aorta which results in stenosis, thrombosis and aneurysm formation. Half of patients present with an initial systemic illness, whereas the other 50% present with late-phase complications.

Two phases of the disease are classically described:
- Pre-pulseless phase: Characterised by nonspecific systemic symptoms
- Pulseless phase: Presents with limb ischaemia or renovascular hypertension

Location and Classification

It has been classified based on location:[3]
- *Type I:* Classic type involving solely the aortic arch branches: Brachiocephalic trunk, carotid and subclavian arteries
- *Type II:*
    - IIa: Involvement of the aorta solely at its ascending portion and/or at the aortic arch +/– branches of the aortic
    - IIb: Involvement of the descending thoracic aorta +/– ascending or aortic arch + branches
- *Type III:* Involvement of the thoracic and abdominal aorta distal to the arch and its major branches, e.g. descending thoracic aorta + abdominal aorta +/– renal arteries
- *Type IV:* Sole involvement of the abdominal aorta and/or the renal arteries
- *Type V:* Generalised involvement of all aortic segments.

Suggested Reading

1. Hata A, Noda M, Moriwaki R, Numano F. Angiographic findings of Takayasu arteritis: new classification. Int J Cardiol 1996; 54 Suppl. S155–63.
2. Nastri MV, Baptista LP, Baroni RH, Blasbalg R, de Avila LF, Leite CC, et al. Gadolinium-enhanced three-dimensional MR angiography of Takayasu arteritis. Radiographics 2004; 24: 773–8.

Case 15

Patient 53-year-old female with complaints of breathlessness, cough for a few days. Patient had some skin reaction on her right arm. So non-contrast MRI and NCCT chest were done to rule out pulmonary embolism or any other pathology.

NONCONTRAST MRI AND NCCT CHEST FINDINGS

- Normal main, right and left pulmonary arteries and normal pressure gradients across them (Figs 12.63 and 12.65).
- No obvious filling defect in pulmonary artery.
- B/L pleural effusion and basal atelectasis with small patchy consolidation (Fig. 12.64).
- Normal LV, RV with normal systolic function (Figs 12.66 and 12.67).

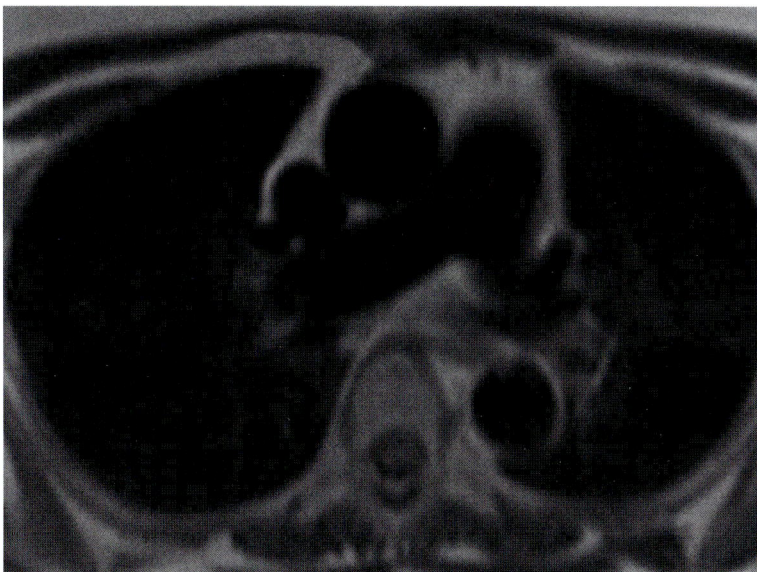

Fig. 12.63: Non-contrast MRI showing RT/LT pulmonary artery, upper, lower branches

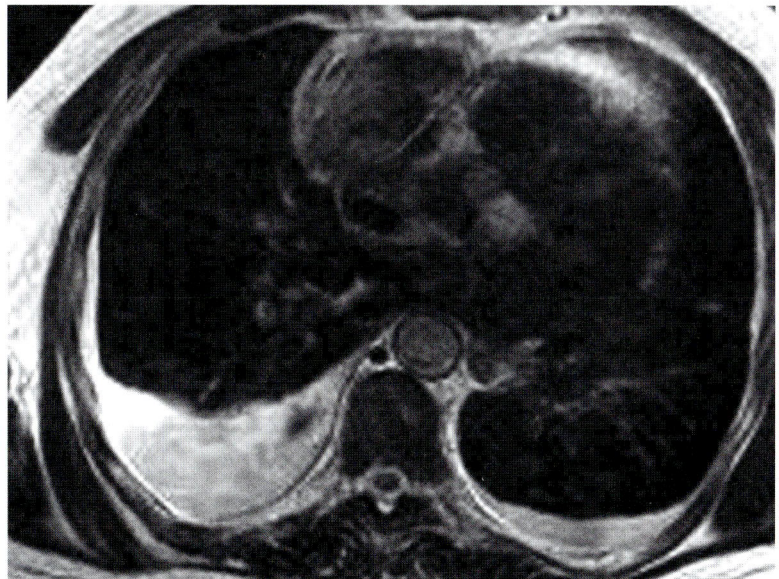

Fig. 12.64: Non-contrast MRI showing basal atelectasis and pleural effusion

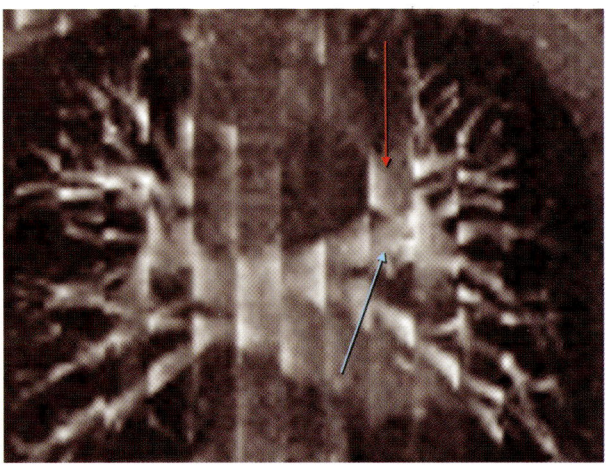

Fig. 12.65: Noncontrast MRI showing normal pulmonary branches (arrow)

Phase-contrast MRI (PC-MRI) was done to evaluate pressure gradient across pulmonary arteries. The pressure gradients were normal.

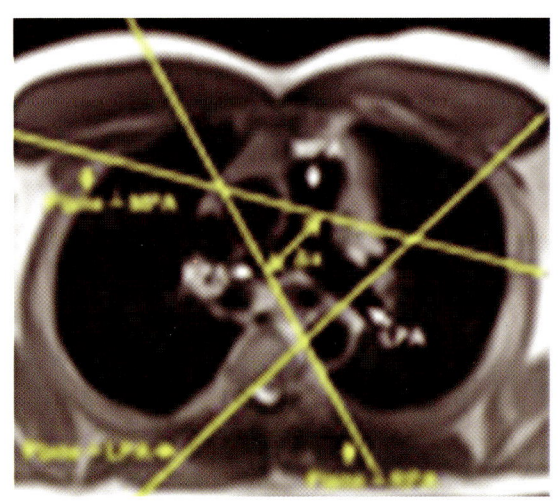

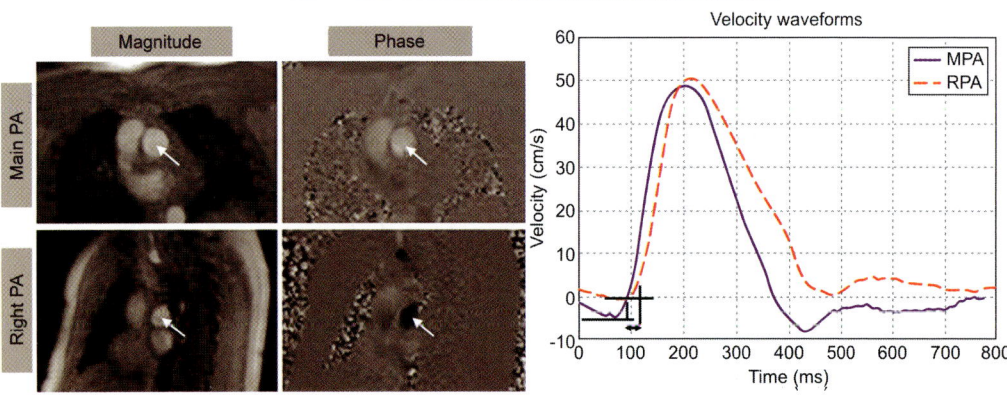

Fig. 12.66: Venc study for pulmonary artery gradients

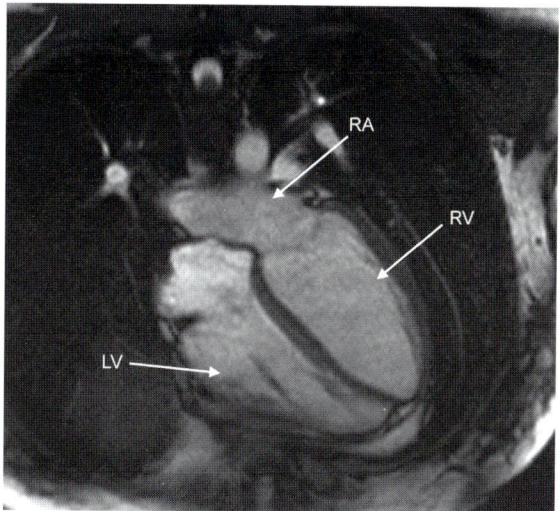

Fig. 12.67: Cardiac MRI showing normal LV and RT ventricle (arrows)

Cardiac MRI was done for left ventricular morphology, motion abnormalities, viability and function evaluation.

Coroborative CT scan done showed B/L pleural effusion, basal atelectasis and small patchy consolidation.

Radiological Diagnosis

Normal main, right and left pulmonary arteries with normal pressure gradients across them, no obvious filling defect in pulm artery suggesting pulmonary thrombolism, normal LV, RV with normal systolic function, B/L pleural effusion and basal atelectasis with small patchy consolidation.

Conclusion

Noncontrast MRA is a noninvasive technique with which flow can be measured accurately with flexible spatial and temporal resolution has a variety of applications in quantifying cardiovascular function and hemodynamics. Compared with contrast-enhanced angiographic sequences, unenhanced sequences demonstrate lower sensitivity, except for proximal PE, but high specificity and agreement.

Suggested Reading

Fink C, Ley S, Kroeker R, Requardt M, Kauczor HU, Bock M. Time-resolved contrast-enhanced three-dimensional magnetic resonance angiography of the chest: combination of parallel imaging with view sharing (TREAT). Invest Radiol 2005; 40: 40–8.

Case 16

A 5-year-old male child presented with complaints of long history of:
- Bilateral axillary swellings and
- Repeated respiratory tract infections.

FNAC was done in his country from the axillary swelling. No records available.

O/E: They appeared as pulsatile swellings.

CT Angiography and Doppler USG Findings

- Left coronary artery aneurysm (Figs 12.68, 12.69 and 12.71a), volume rendering image
- Ectasia and stenosis of left subclavian artery and left axillary artery aneurysm (Figs 12.70a and b)
- Heart and the rest of great vessels showed no abnormality
- Bilateral lung fields were clear
- Doppler USG showed consistent findings (Figs 12.17b, c, 12.72, 12.73a and b).

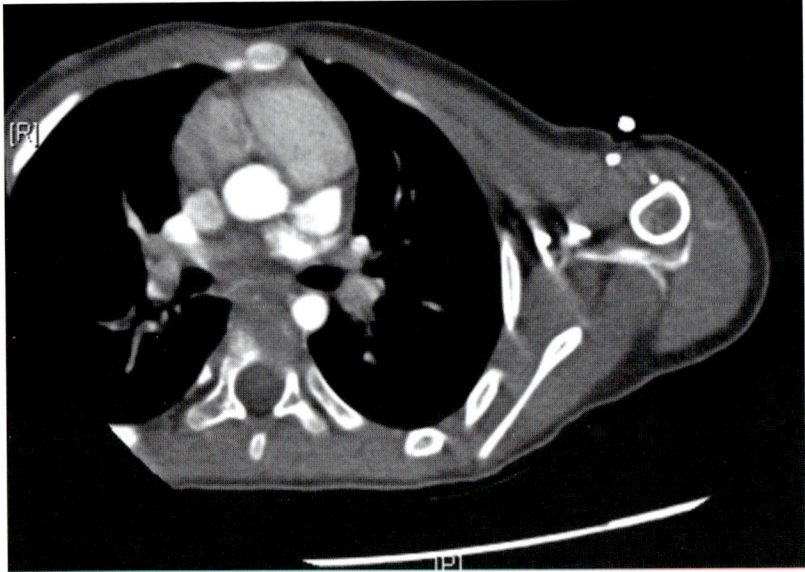

Fig. 12.68: CT angiography showing aneurysm of proximal left main coronary artery

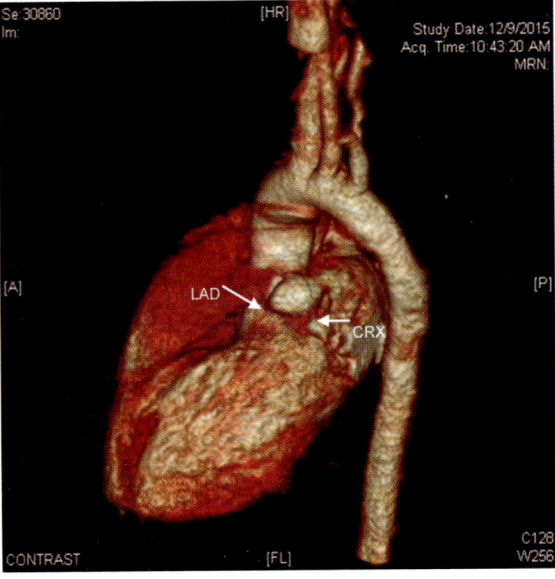

Fig. 12.69: CT volume rendering image

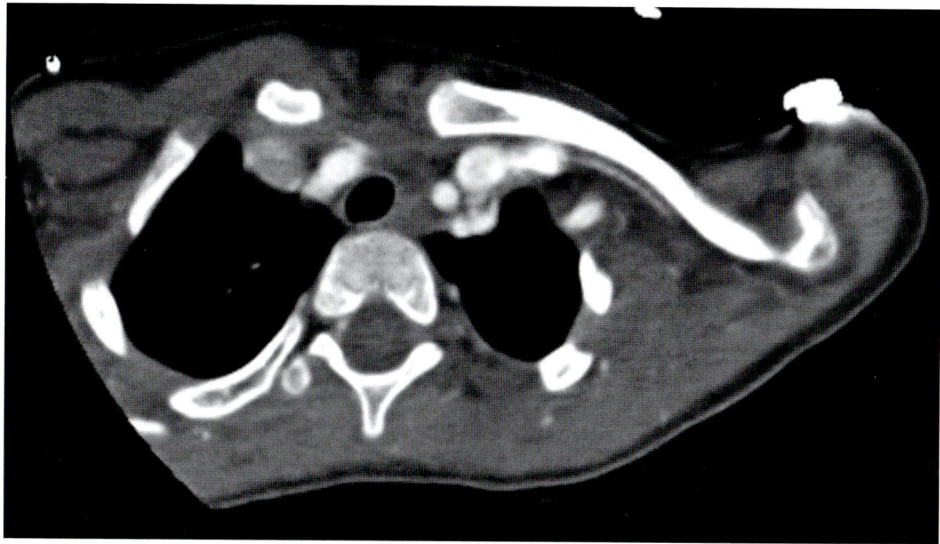

(a)

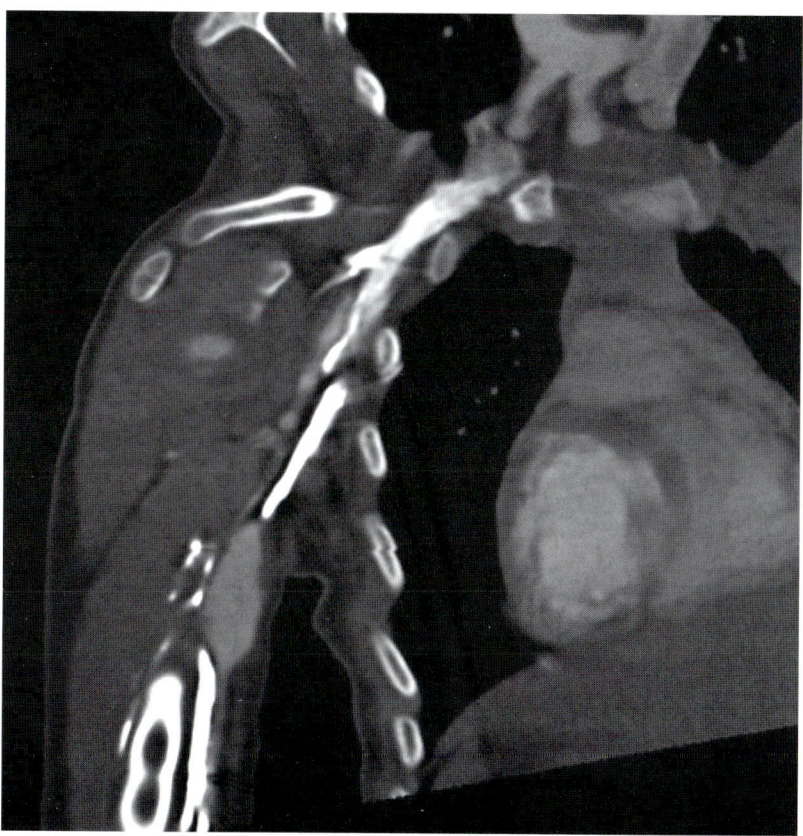

(b)

Fig. 12.70a and b: CT showing left subclavian and axillary artery aneurysm

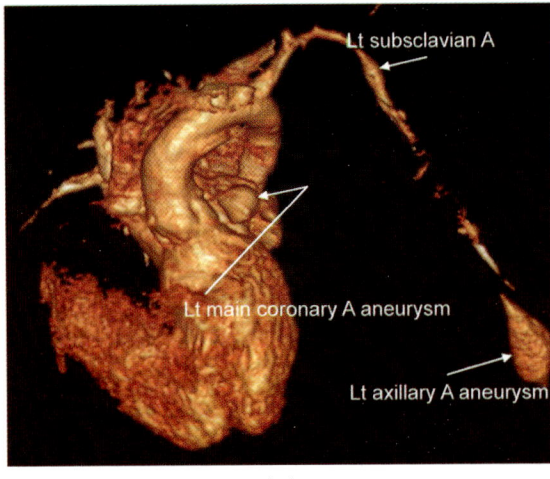

(a)

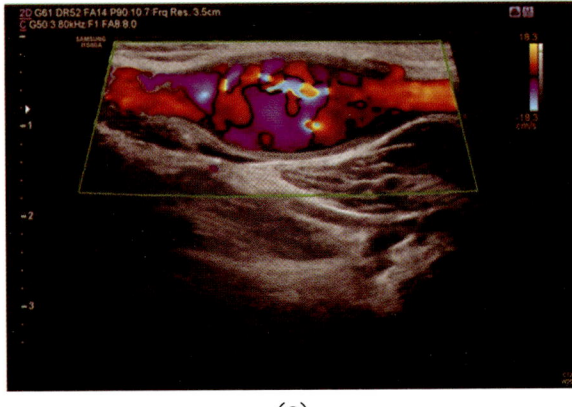

(b)

(c)

Fig. 12.71a, b and c: CT volume rendering image and Doppler showing left SCA and axillary artery aneurysm

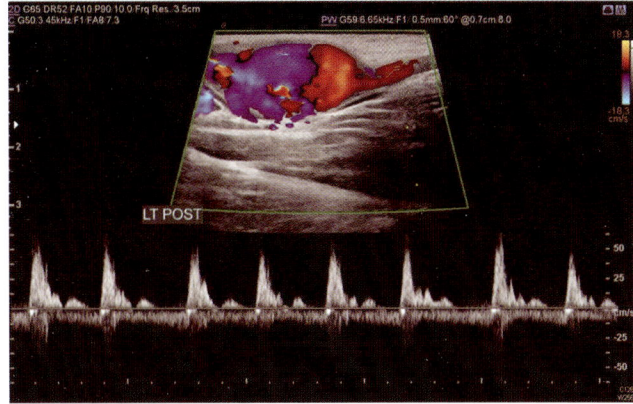

Fig. 12.72: Doppler USG distal to axillary artery aneurysm

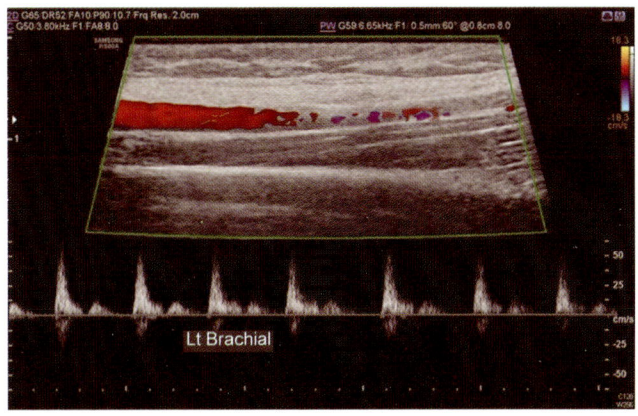

(a)

(b)

Fig. 12.73a and b: Doppler showing normal flow in left brachial and ulnar artery

Discussion

- The rest of the coronary arteries, heart and the great vessels showed no abnormality
- Bilateral lung fields were clear.
- The features of coronary artery aneurysm with ectasia and stenosis of left subclavian artery and left axillary artery aneurysm in a 5-year-old child with history of repeated respiratory tract infections suggest a possibility of small to medium vessel vasculitis, likely Kawasaki disease.

Suggested Reading

1. Images in cardiovascular medicine. Evaluation of coronary artery aneurysms in Kawasaki disease by multislice computed tomographic coronary angiography. Wu MT, Hsieh KS, Lin CC, Yang CF, Pan HB. Circulation. 2004 Oct 5; 110(14): e339.
2. Assessment of coronary artery abnormalities in a patient with Kawasaki disease by multislice computed tomography. Sato Y, Matsumoto N, Inoue F, Imazeki T, Kusama J, Tamaki T, Furuhashi S, Takahashi M, Kanamaru H, Karasawa K, Ayusawa M, Harada K, Kanmatsuse K. Heart Vessels. 2004 Nov; 19(6): 297.

Case 17

Patient is 45-year-old developed bleeding from endotracheal tube. Not fit for rigid bronchoscopy to assess bleeding site. CT angiography was advised.

CT Angiography Findings

1. Focal vessel injury of RT brachiocephalic artery near its bifurcation with extravasation of contrast seen around the tube (Figs 12.74 and 12.75).
2. Normal calibre of RCCA, RVA and LVA and LCCA.
3. Multiple areas of consolidation and centrilobular nodules in bilateral lung fields (Fig. 12.80).

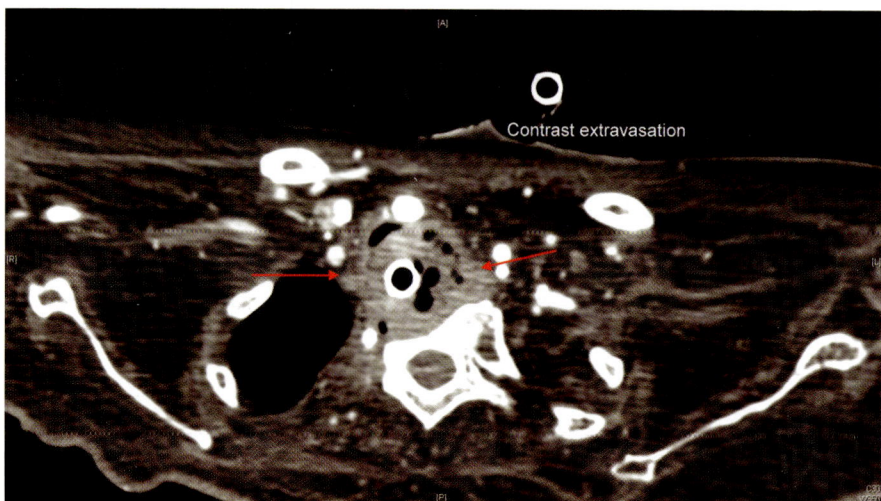

Fig. 12.74: CT showing extravasation of contrast surrounding endotracheal tube

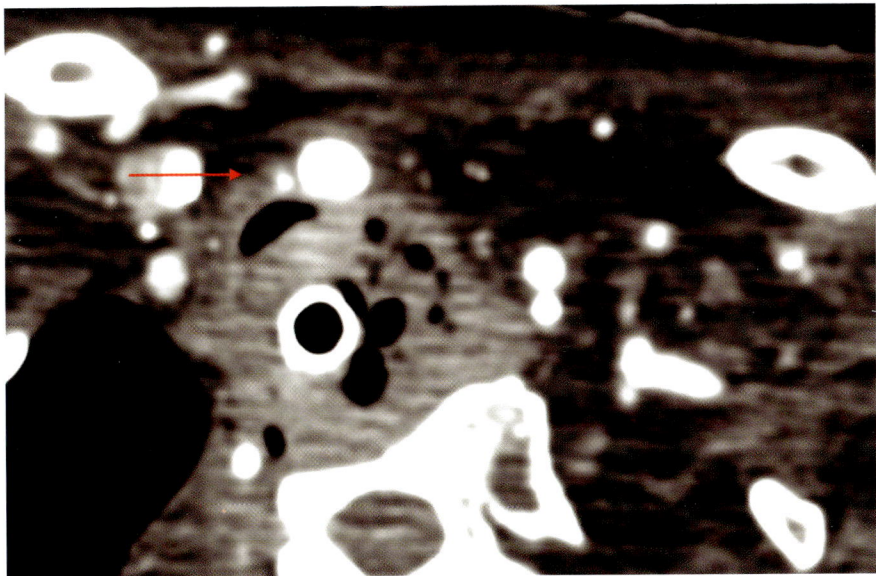

Fig. 12.75: CT showing injury just distal to RT innominate bifurcation. RT SCA and RT innominate artery at site of injury

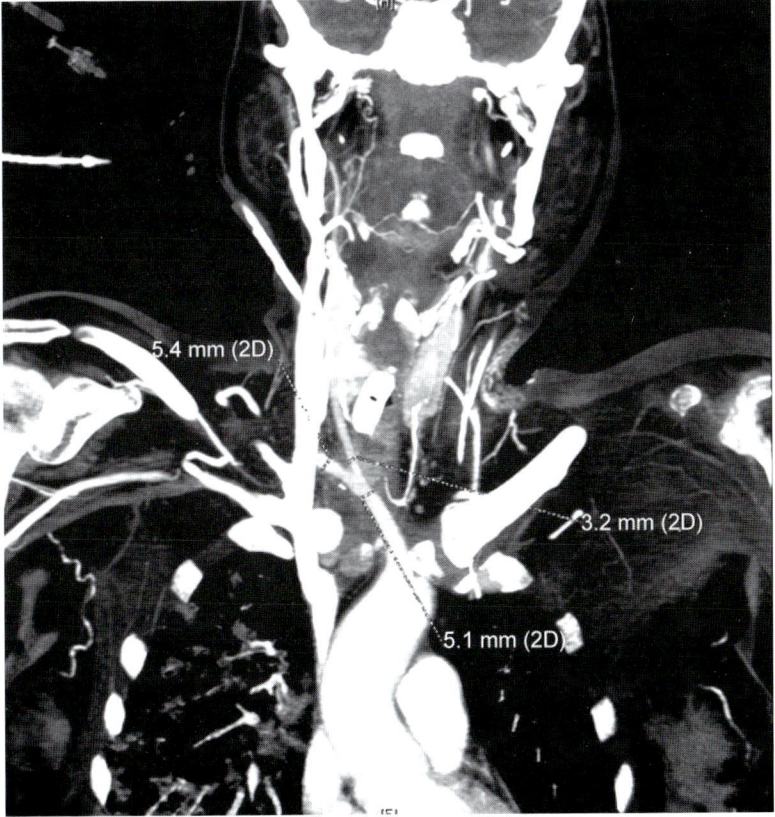

Fig. 12.76: CT showing normal calibre of RCCA

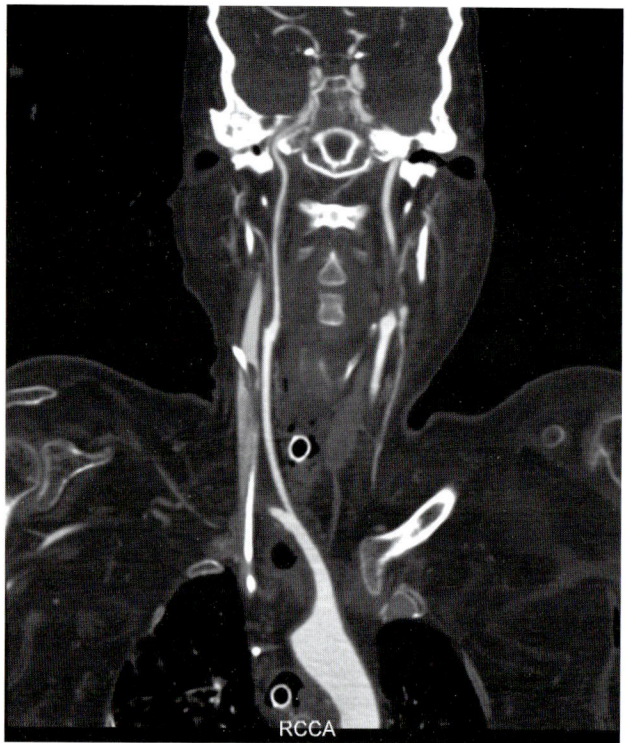

Fig. 12.77: CT showing RVA

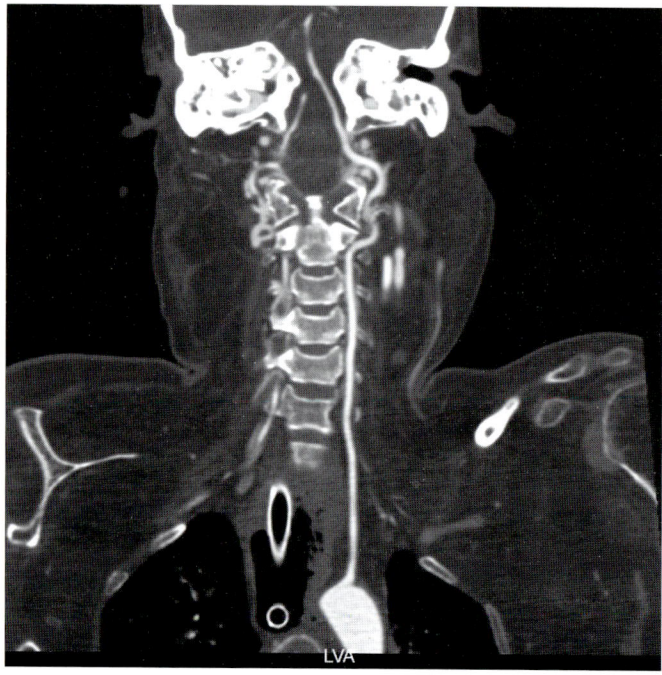

Fig. 12.78: CT showing normal calibre LVA

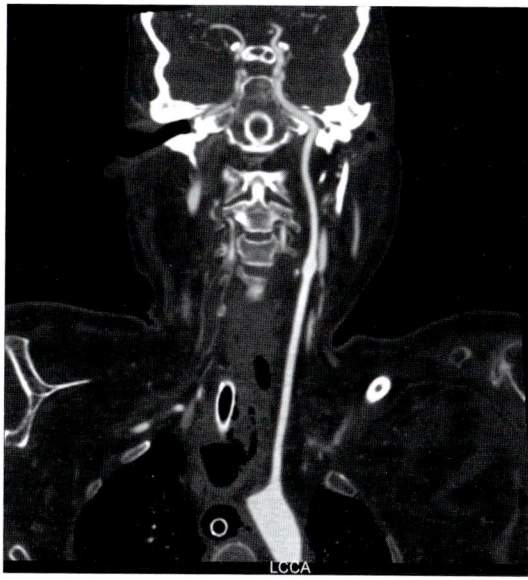

Fig. 12.79: CT showing normal calibre of LCCA

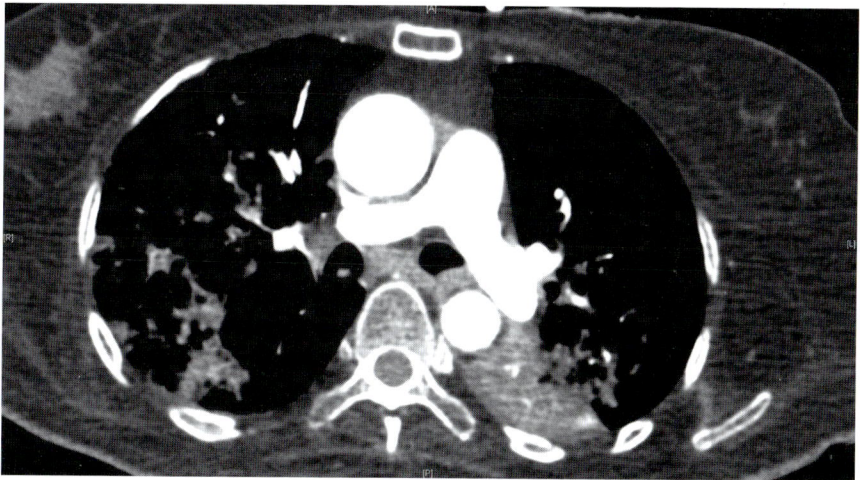

Fig. 12.80: CT showing B/L lower lobe consolidation

Radiological Diagnosis

Focal vessel injury of RT brachiocephalic artery near its bifurcation and multiple areas of consolidation and centrilobular nodules in bilateral lung fields.

Surgical Intervention

- Procedure PTA with stents (covered stents: 6 × 60 mm SCA, 6 × 40 mm CCA) to right CCA and SCA (chimney) from right CFA and right brachial artery route under GA (Figs 12.81 and 12.82).
- Covered stent placement in the innominate artery, as a life saving measure (Figs 12.83 and 12.84)

Surgical Intervention Findings Were

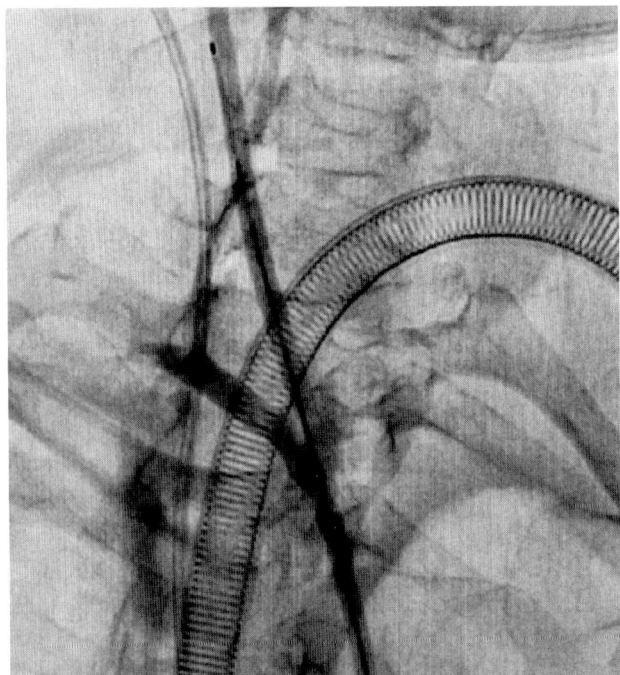

Fig. 12.81: Extravasation of contrast from CCA blowout

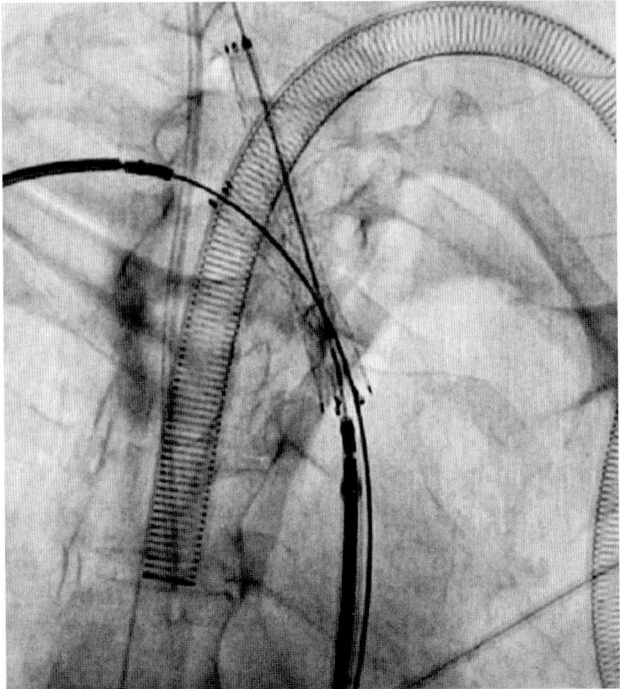

Fig. 12.82: Deployment of double barrel stents into CCA and subclavian artery

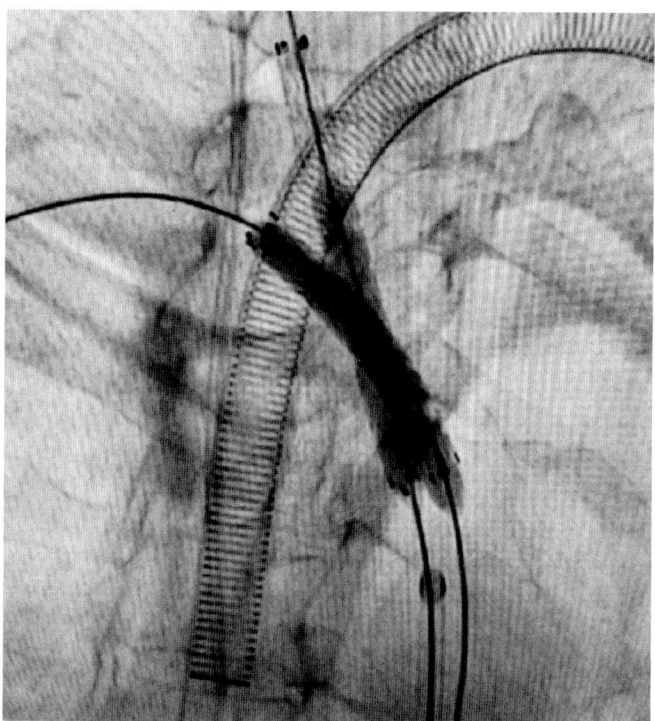

Fig. 12.83: Ballooning of covering stents

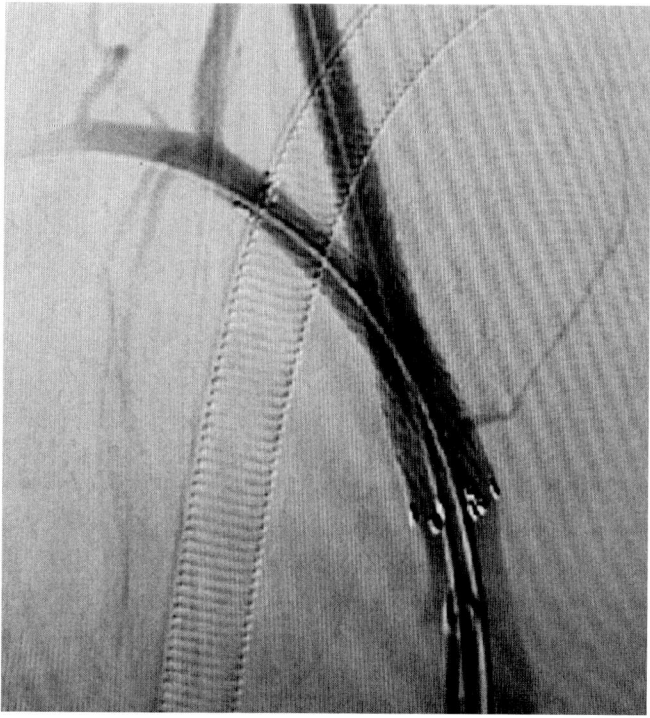

Fig. 12.84: Final run with no extravasation of contrast

Conclusion

CTA is an important diagnostic imaging modality for the evaluation of upper extremity arterial abnormalities. High-quality CTA of the upper extremities is feasible on modern CT scanners using optimized scanning and contrast injection technique. Its 24-hour availability, rapid acquisition, minimal invasiveness, and display of both vascular and musculoskeletal structures makes it particularly attractive for the evaluation of patients with blunt or penetrating trauma to the upper extremity.

Suggested Reading

1. Hellinger JC, Epelman M, Rubin GD. Upper extremity computed tomographic angiography: state of the art technique and applications in 2010. Radiol Clin North Am 2010; 48: 397–421.
2. Chait A. Arteriography of the upper extremity. In: Baum S, ed. Abrams' angiography. New York, NY: Little, Brown and Company, 1997: 1755–1767.

Case 18

- A 65-year-old female, a known case of bronchial asthma 7–8 years.
- Presented with c/o epigastric pain radiating to back for 4–5 days
- Pain relieved slightly after eating and reoccurs when the stomach is empty
- No vomiting/loose motions, no resp complaints.
- ECG/chest X-ray: No significant abnormality.
- CT angiography was done for further evaluation.

CT Angiography Findings Were

- Partially thrombosed aneurysmal dilatation of the thoracic descending aorta measuring 5.9 × 5.5 × 5.5 cm in size with patent lumen measures 4.4 cm (Figs 12.85, 12.86 and 12.87).
- A few atheromatous ulcers are seen in the abdominal aorta distal to aneurysm.
- Emphysematous changes in lung fields
- Fibrotic opacity bilaterally in the lungs.
- Subcentimeter mediastinal lymph nodes.

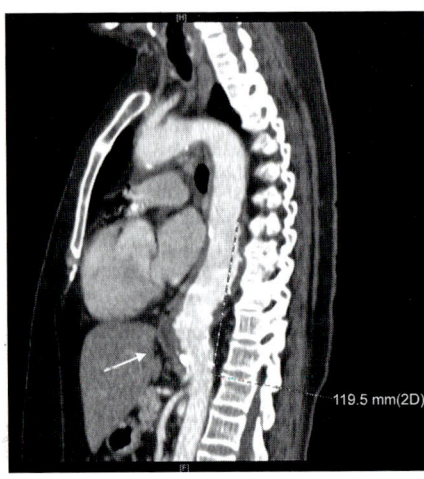

Fig. 12.85: Partially thrombosed aneurysmal dilatation of the descending aorta (arrow)

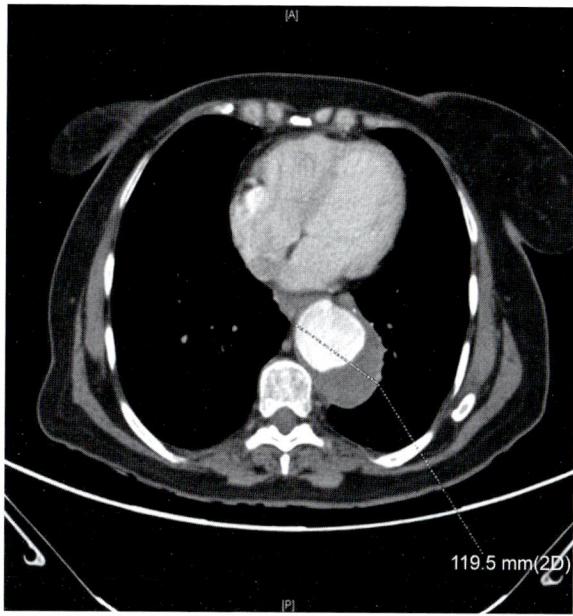

Fig. 12.86: CT angiography showing peripheral thrombus

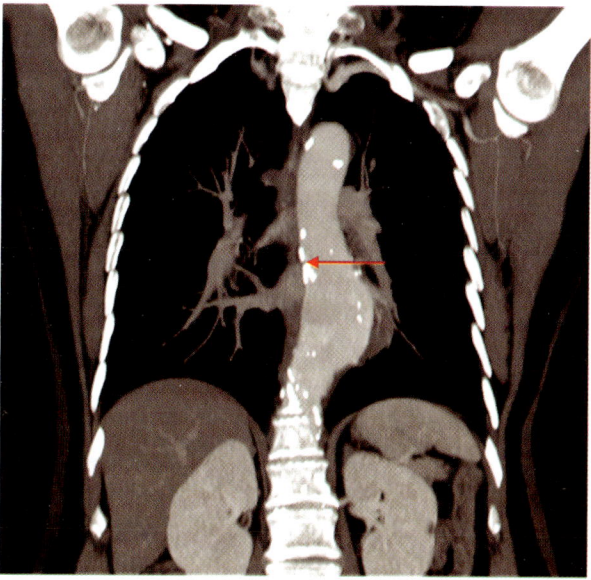

Fig. 12.87: CT showing partially thrombosed aneurysmal dilatation of the descending aorta

Radiological Diagnosis

Partially thrombosed aneurysmal dilatation of the descending with a few atheromatous ulcers are seen in the abdominal aorta distal to aneurysm.

Surgical Intervention

Stent graft was deployed. Further angiography showed no endoleak (Figs 12.88 and 12.89).

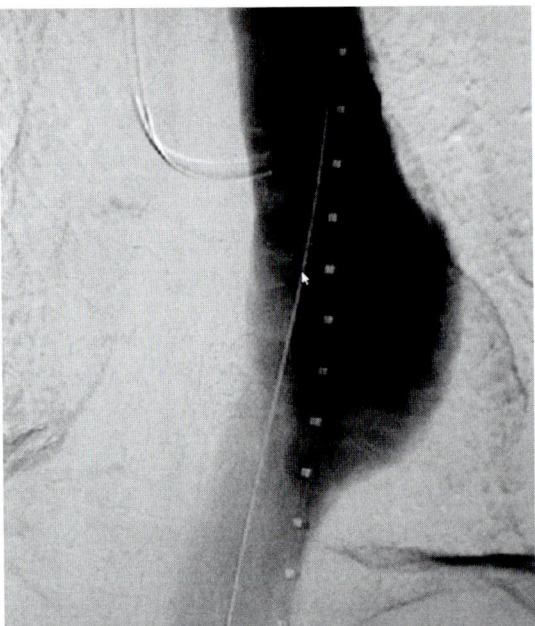

Fig. 12.88: Conventional angiography showing descending aortic aneurysm before the surgical intervention

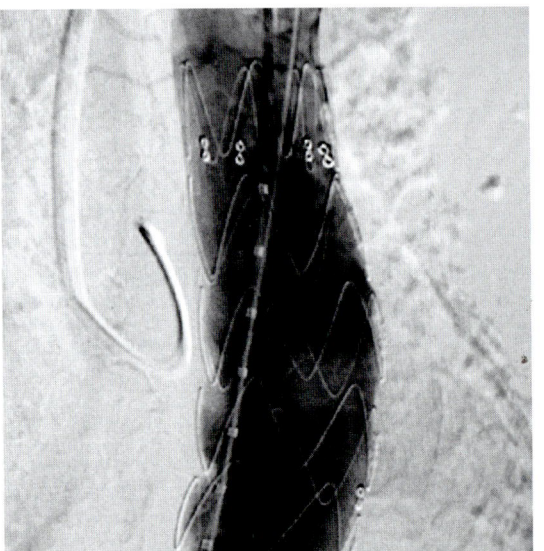

Fig. 12.89: Stent graft in descending aorta

Conclusion

Whereas aortic aneurysms are less common than many other cardiovascular conditions, the fact that they can be life threatening and that even large aneurysms may not produce symptoms makes it all the more important for clinicians to be vigilant in their evaluation of patients at risk. Because aneurysms are often first detected on an imaging study ordered for other indications, any suggestion of an enlarged aorta should prompt follow-up with an appropriate dedicated imaging study. Fortunately, modern imaging techniques—especially CT and MRI—have now made the sizing and surveillance of aneurysms relatively easy. In the future, genetic screening may also play a role in the screening of those with a family history of thoracic aortic aneurysms. Ideally, a broadening clinical awareness of aortic aneurysms and the methods of diagnosis will help reduce the morbidity and mortality associated with this condition. Indeed, clinicians who understand the principles of medical and surgical management of aortic aneurysms can comfortably determine when they should manage patients with medication and serial imaging studies (to follow aneurysm size and rate of growth) and when to refer to a cardiothoracic or vascular surgeon. Finally, whereas open surgical repair remains the standard approach to treating most large aortic aneurysms, it is likely that endovascular stent-grafting will assume an increasingly important role as the technique is further refined.

Suggested Reading

1. Guo D, Hasham S, Kuang S-Q, Vaughan CJ, Boerwinkle E, Chen H, Abuelo D, Dietz HC, Basson CT, Shete SS, Milewicz DM. Familial thoracic aortic aneurysms and dissections. Circulation. 2001; 103: 2461–2468.
2. Jeremy RW, Huang H, Hwa J, McCarron H, Hughes CF, Richards JG. Relation between age, arterial distensibility, and aortic dilatation in the Marfan syndrome. Am J Cardiol. 1994; 74: 369–373.

Case 19

- Patient has history of recurrent respiratory tract infections
- No h/o of diarrhoea/vomiting.
- No h/o of poor weight gain.
- No h/o recurrent hospital admissions.

Echocardiography and CT scan were done.

Echocardiography Findings

- Bicuspid aortic valve
- Severe AS Peak PG = 147 mm of Hg
- Mean Pg = 87 mm of Hg
- Severe juxtaductal coarctation of aorta (discrete)
- Coarct gradient = 29 (?pandiastolic split), trivial AR
- Hypertrophic LV
- AO annulus = 11.5 mm
- Sinus = 17 mm
- LVEF = 50–55%

CT ANGIOGRAPHY FINDINGS

- Normal left coronary artery. Juxtaductal coarctation of aorta with aberrant right subclavian artery originating distal to it. Rest of aorta was normal (Figs 12.91, 12.92, 12.93 and 12.94)
- No PDA noted.

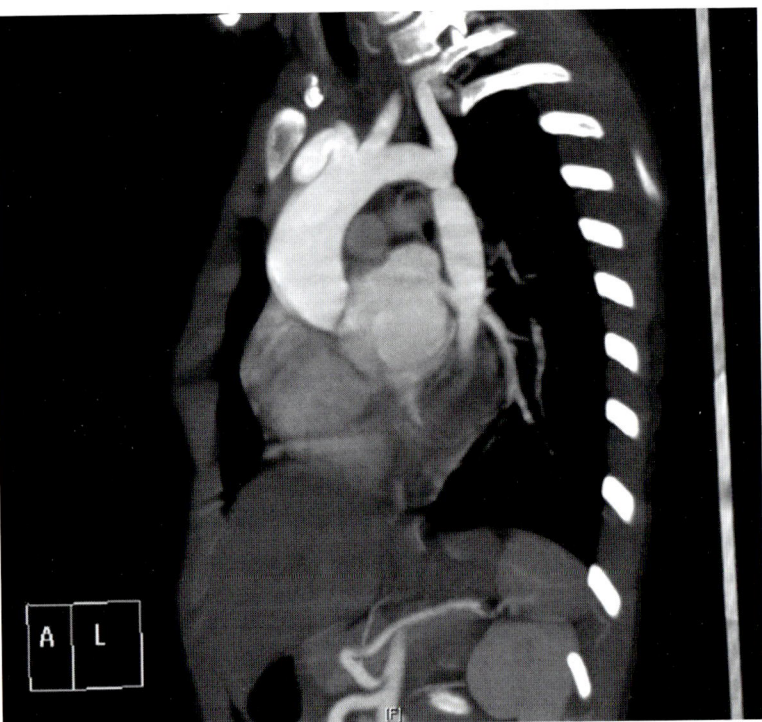

Fig. 12.90: CT showing coarctation distal to origin of left SCA

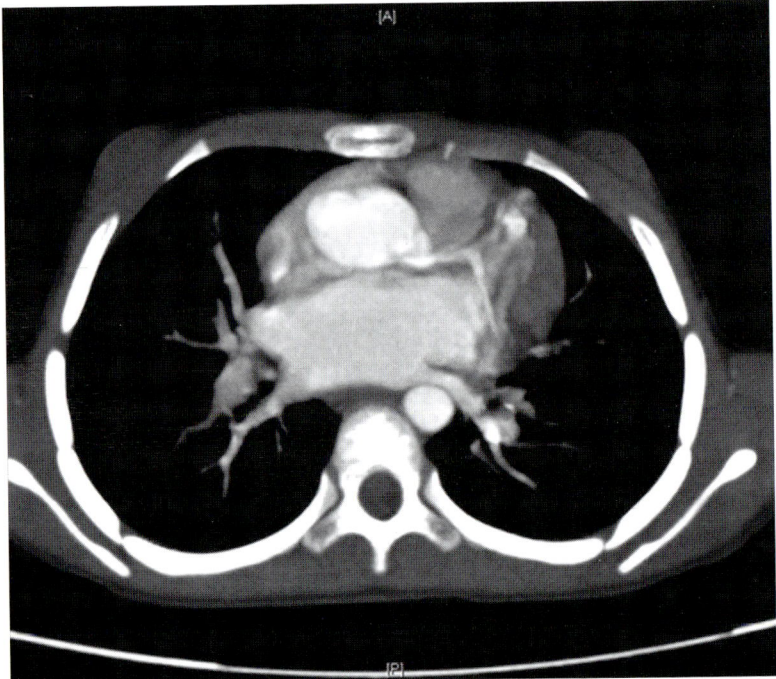

Fig. 12.91: CT showing normal origin of left main coronary artery

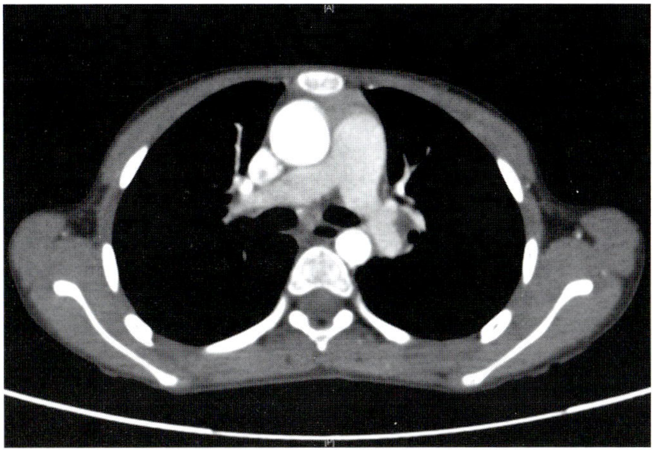

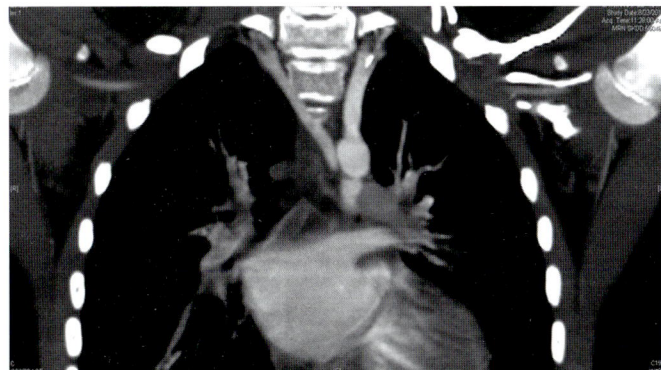

Fig. 12.92: CT showing ascending and descending aorta

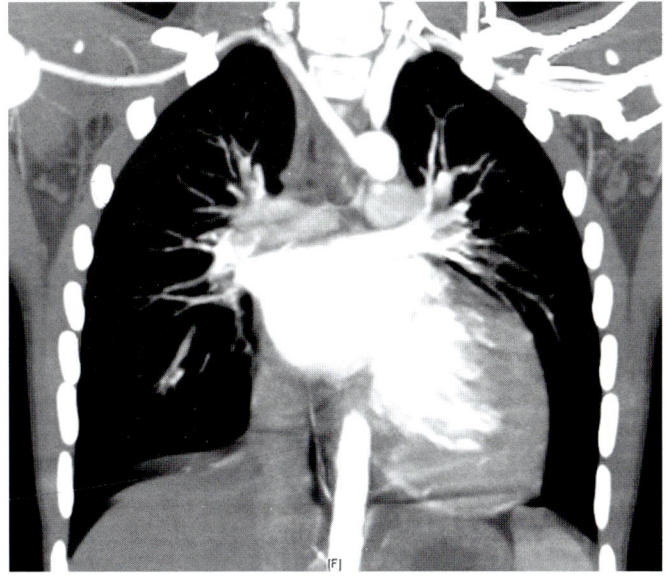

Fig. 12.93: CT coronal reformatted image showing coarctation segment

Fig. 12.94: CT showing aberrant origin of RT subclavian artery

Radiological Diagnosis

Juxtaductal coarctation of aorta with aberrant subclavian artery originating distal to it and no PDA was noted.

Surgical Intervention

Aortic valve placement and coarctation ballooning were done.

Conclusion

Aortic coarctation is one of the most common congenital heart diseases, and early diagnosis and treatment are the keys to successful outcomes. Cardiac echocardiography is generally the first imaging test because of its ease of use and lack of ionizing radiation. However, not all segments of the aorta can be optimally evaluated with this modality, and CT or MRI is almost always used for surgical planning and follow-up.

With the introduction of modern noninvasive imaging modalities, the diagnostic use of conventional catheter angiography has almost completely disappeared. CT and MRI can provide excellent anatomic detail as well as direct and indirect functional data. With the use of these cross-sectional modalities, the surgical and medical care teams can be provided with accurate information for surgical planning and follow-up in a completely noninvasive manner. Particularly with the functional information provided with phase contrast imaging, the pressure gradient across the coarctation segment, which is an important indicator during the decision-making process, can be accurately predicted. Another important advantage of these noninvasive imaging modalities is simultaneous evaluation of the heart for associated congenital defects.

With modern scanning techniques and state-of-the-art CT scanners, the overall radiation burden on the patients has also been significantly reduced; however, considering the need for lifelong follow-up of young patients, the radiologist should still be prudent and cautious with the use of modalities using ionizing radiation. Therefore, MRI may be a better option for these patients. Local expertise and availability of the modalities are important parameters that should be taken into account before making the final decision.

Suggested Reading

1. Kimura-Hayama ET, Meléndez G, Mendizábal AL, Meave-González A, Zambrana GF, Corona-Villalobos CP. Uncommon congenital and acquired aortic diseases: role of multidetector CT angiography. RadioGraphics 2010; 30: 79–98.
2. Becker AE, Becker MJ, Edwards JE. Anomalies associated with coarctation of aorta: particular reference to infancy. Circulation 1970; 41: 1067–1075.

Case 20

- A 16-year-old female admitted with h/o CHF following episode of fever for last 1 year, worsened in last three weeks (NYHA class iii symptoms)
- Pedal edema progressed in last 1 month
- H/O taking ATT for pulmonary Koch's (no documents available) for 2 months stopped in outside hospital because of hepatitis.
- On examination: Sensorium: dull, oriented, no focal deficits

- USG abdomen (14.11.15) reveals bilateral pleural effusion (moderate) with segmental collapse. Moderate ascites. Small left kidney with raised echogenicity and poor CMD. Corticomedullary differentiation borderline hepatomegaly.
- Plan
- To try systemic steroids and see response in LV function and heart rate and features of heart failure.

Echocardiography was Done

Transthoracic echo-Doppler was done which showed normal segmental analysis. Biventricular dysfunction. LVEF~ 15%. Trivial MR. Mild TR (max PG~51 mmHg). Mild PR (max PG~ 23 mmHg). Laminar flow in LVOT/RVOT. Normal arch. No COA. Normal coronaries. B/L pleural collection. Moderate pericardial collection all around the heart. No RADC/RVDC. IVC size ~ 18 mm.

CT Angiography was Done Which Showed

- Cardiomegaly with left ventricular chamber enlargement (Fig. 12.95)
- Pericardial and pleural effusion (Fig. 12.95)
- Ascitis, segmental irregularity and narrowing of the descending thoracic and upper abdominal aorta with attenuated calibre of celiac trunk origin and proximal part of its branches (Figs 12.96, 12.97a and b)
- Attenuated left renal artery (Fig. 12.98).
- Ground glassing and air space opacities in bilateral lung fields (Fig. 12.99)

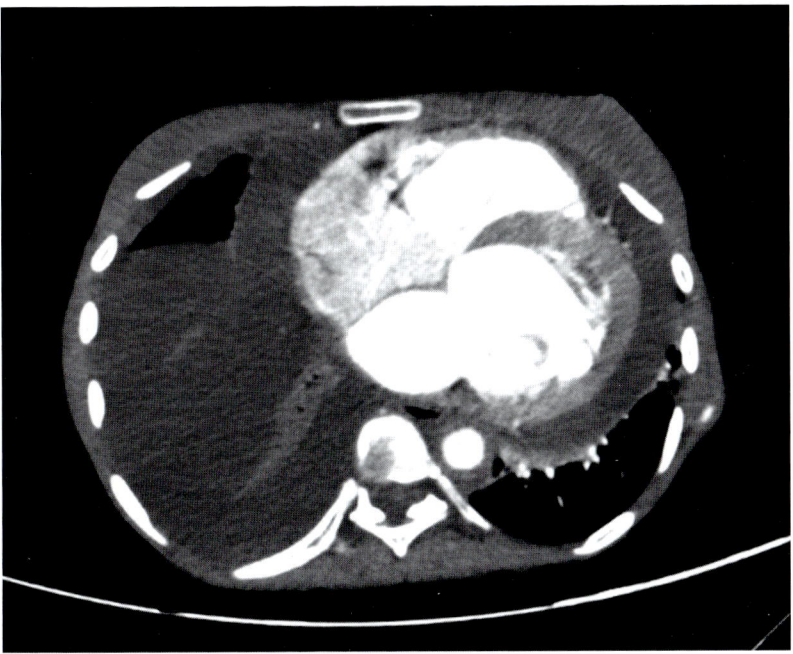

Fig. 12.95: CT showing B/L pleural effusion

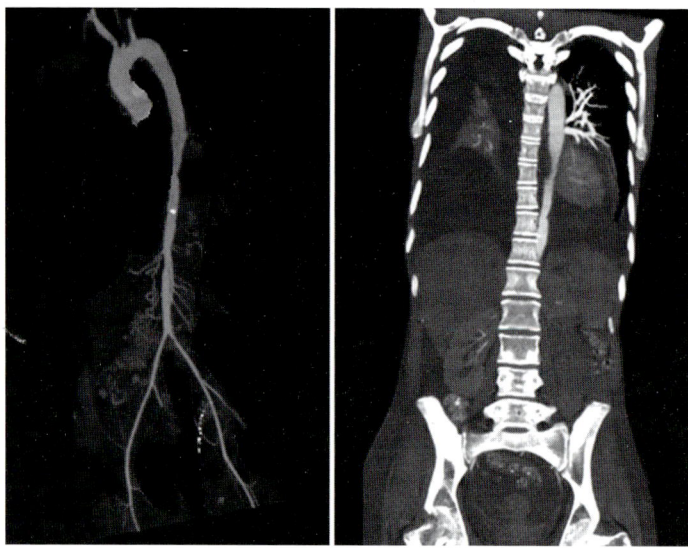

Fig. 12.96: CT showing narrow descending thoracic and abdominal aorta

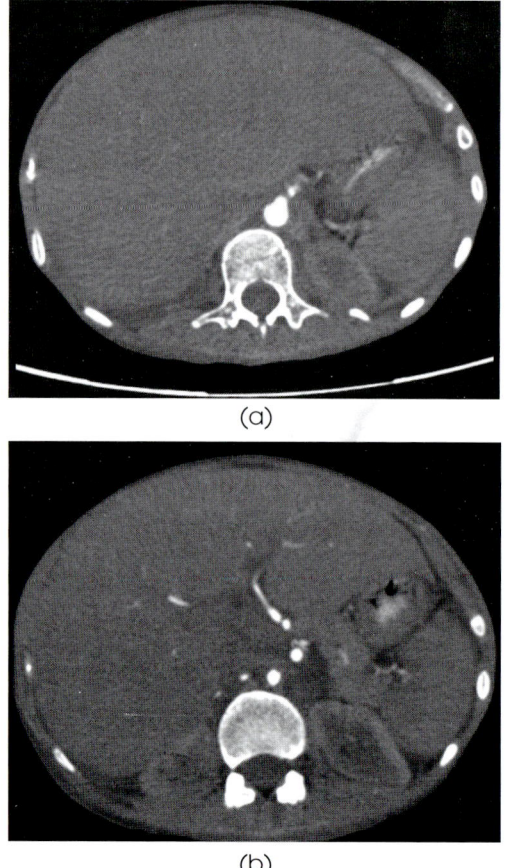

(a)

(b)

Fig. 12.97a and b: CT showing narrow celiac artery and abdominal aorta

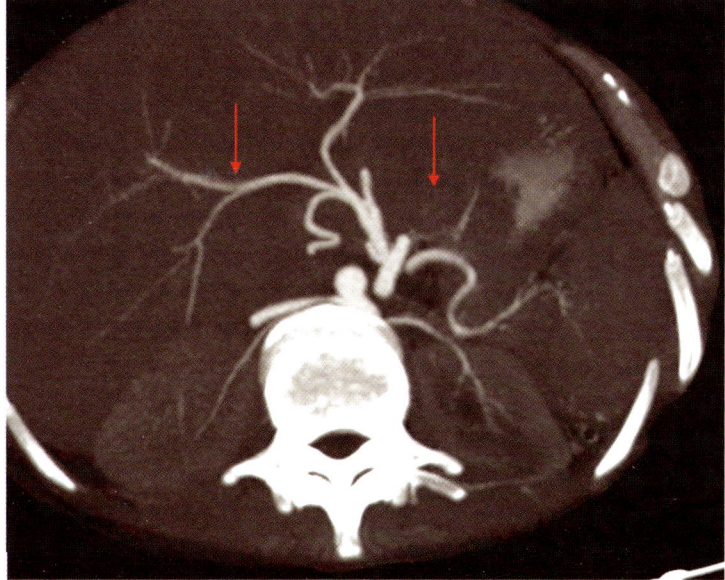

Fig. 12.98: CT showing narrow renal arteries

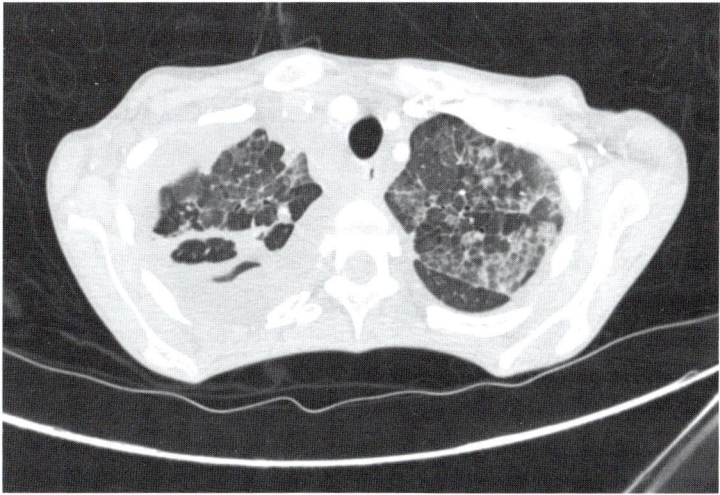

Fig. 12.99: CT showing bilateral ground glass opacity and intralobular septal thickening

Conclusion

Takayasu arteritis (TA) is a rare entity with poor prognosis. Familiarity with its CT appearances can aid the radiologist to make the appropriate diagnosis.

Takayasu Disease

Pathology

Two phases of the disease are classically described:

- Pre-pulseless phase: Characterised by nonspecific systemic symptoms
- Pulseless phase: Presents with limb ischaemia or renovascular hypertension

Location and Classification

It has been classified based on location

- *Type I:* Classic type involving the solely the aortic arch branches: Brachiocephalic trunk, carotid and subclavian arteries
- *Type II:*
    - IIa: Involvement of the aorta solely at its ascending portion and/or at the aortic arch +/– branches of the aortic arch
    - IIb: Involvement of the descending thoracic aorta +/– ascending or aortic arch + branches
- *Type III:* Involvement of the thoracic and abdominal aorta distal to the arch and its major branches, e.g. descending thoracic aorta + abdominal aorta +/– renal arteries
- *Type IV:* Sole involvement of the abdominal aorta and/or the renal arteries
- *Type V:* Generalised involvement of all aortic segments.

Suggested Reading

1. Hata A, Noda M, Moriwaki R, Numano F. Angiographic findings of Takayasu arteritis: new classification. Int J Cardiol 1996; 54 Suppl. S155-63.
2. Canyigit M, Peynircioglu B, Hazirolan T, Dagoglu MG, Cil BE, Haliloglu M, et al. Imaging characteristics of Takayasu arteritis.CardiovascIntervent Radiol 2007; 30: 711–18.

Case 21

Patient's status is post-left brachial artery embolectomy. Now again absent radial pulse in left upper limb. Patient was advised for CT angiography of left upper limb.

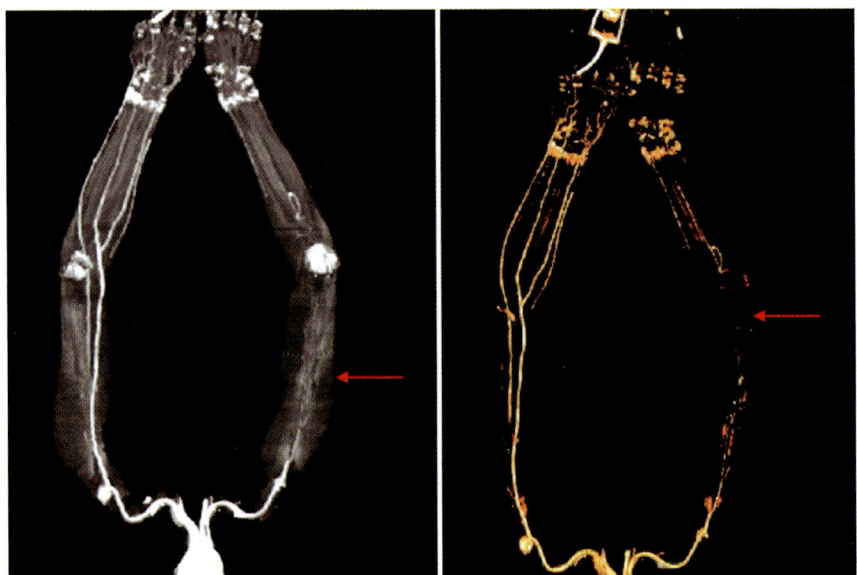

Fig. 12.100: CT angiography showing complete occlusion of distal left brachial artery, severe stenosis of proximal radial and ulnar artery with collaterals supplying the mid and distal radial, ulnar arteries

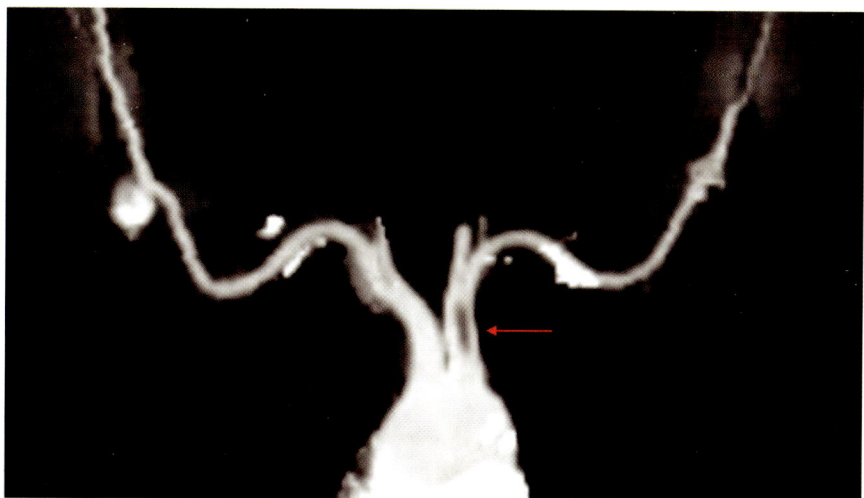

Fig. 12.101: CT angiography showing severe stenosis of proximal left subclavian artery

Ct Angiography Findings

1. Almost complete occlusion of distal left brachial artery (Fig. 12.100).
2. There was severe stenosis of proximal left subclavian, proximal radial and ulnar artery with collaterals supplying the mid and distal radial, ulnar arteries (Figs 12.100 and 12.101).

Surgical Intervention

- Embolectomy of left subclavian artery
- Stenting of left subclavian and brachial arteries.

Case 22

Patient's status is postoperative coarctation of aorta repair and PDA ligation.

Now patient had h/o fever and headache on off for 2 months.

Investigated routinely and USG done which show spenic abscess.

O/E

GPE—mild pallor

CVS—normal

CT Angiography was Planned. The Findings Were

- Lobulated saccular aneurysmal dilatation of desending aorta likely pseudoaneurysms in postoperative region with wall thickening and irregularity (Figs 12.102, 12.105 and 12.106)
- Dilated LV
- Irregular hypodensities in aorta—vegetation in aorta (Fig. 12.103)
- Splenic abscesses (Fig. 12.104)
- Hepatic arterial tortousity in periportal region—precursor of periportal fibrosis.
- Rest of the aorta was normal in calibre (Fig. 12.107)

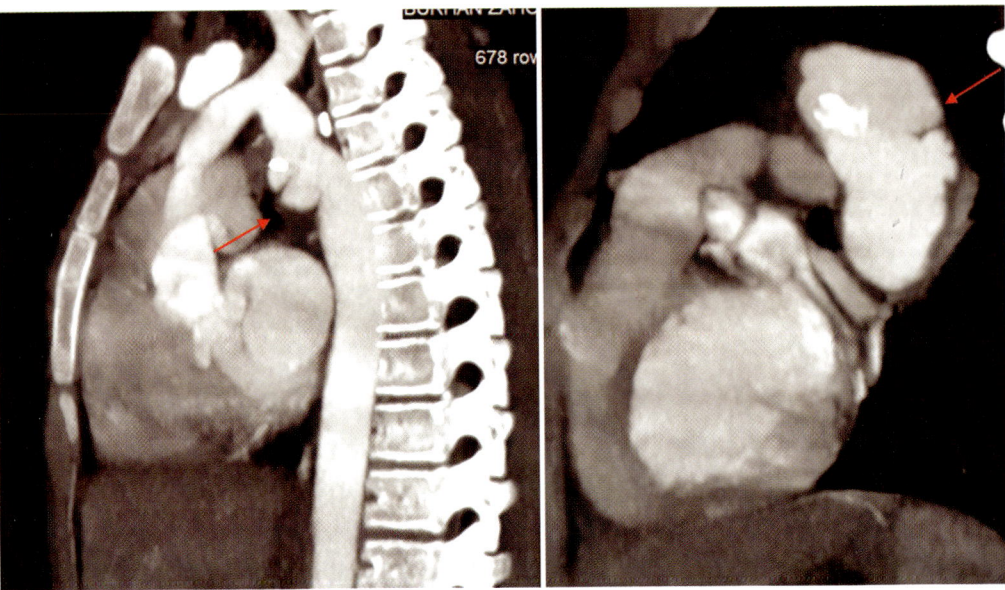

Fig. 12.102: CT angiography showing aortic wall irregularity and outpouching in postoperative region (arrow)

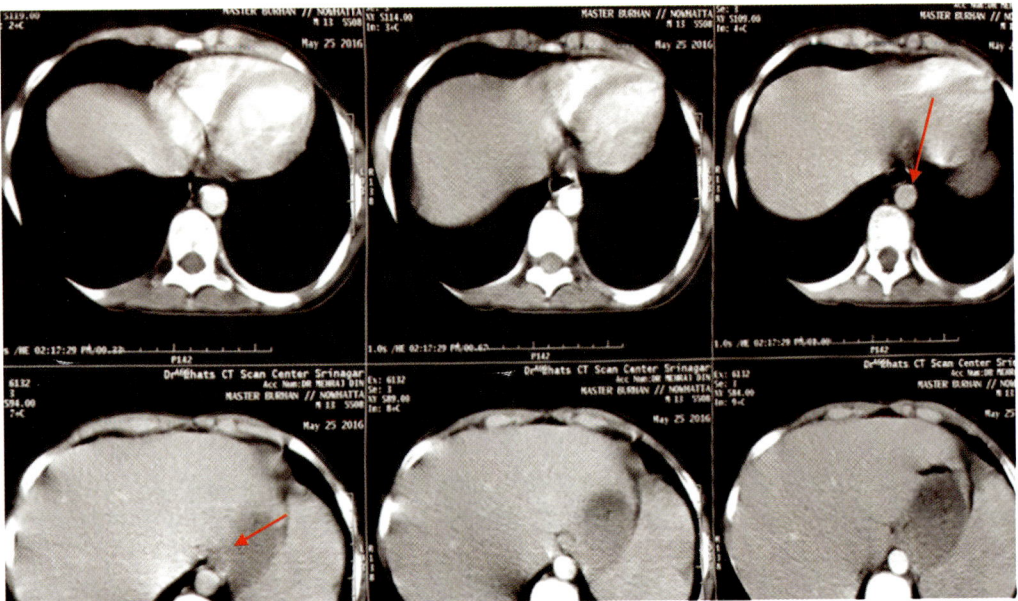

Fig. 12.103: CT showing aortic vegetation (arrow)

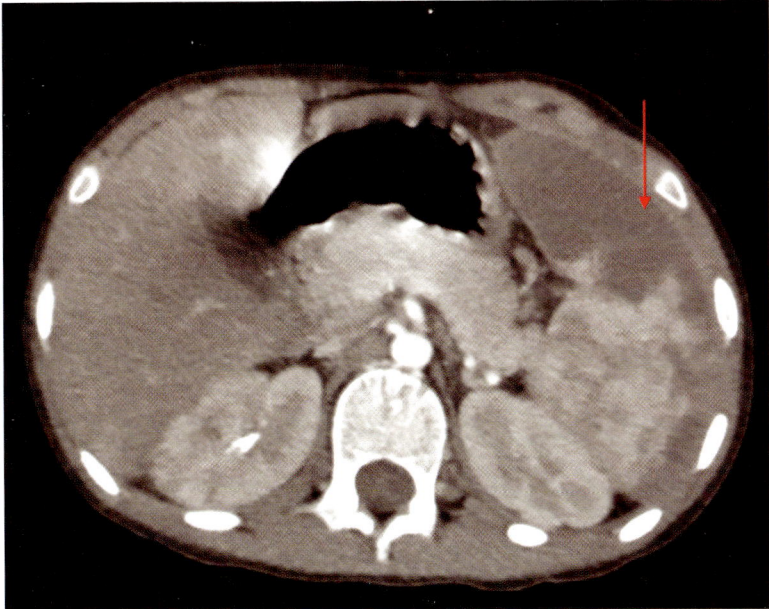

Fig. 12.104: CT scan in abdominal section showing splenic abscess (arrow)

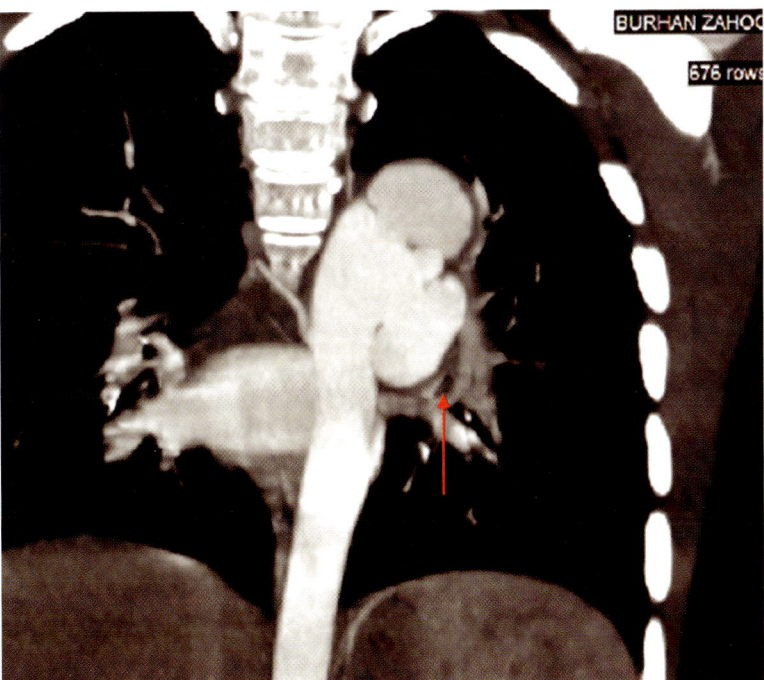

Fig. 12.105: CT angiography showing lobulated pseudoaneurysms arising from descending aorta in operative region (arrow)

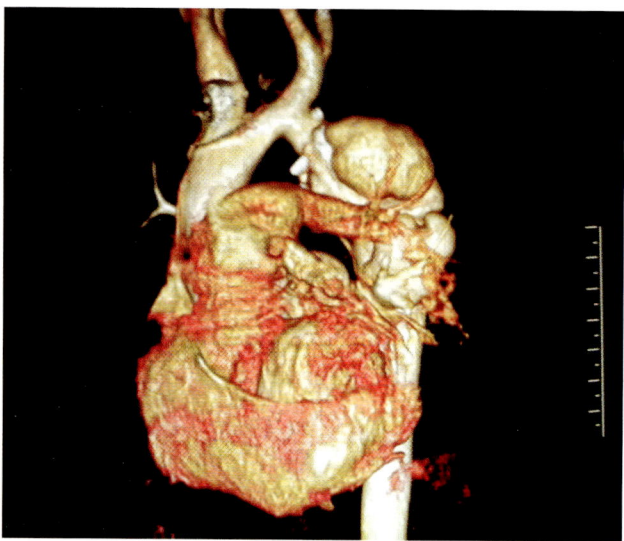

Fig. 12.106: CT volume rendering image showing pseudoaneurysms

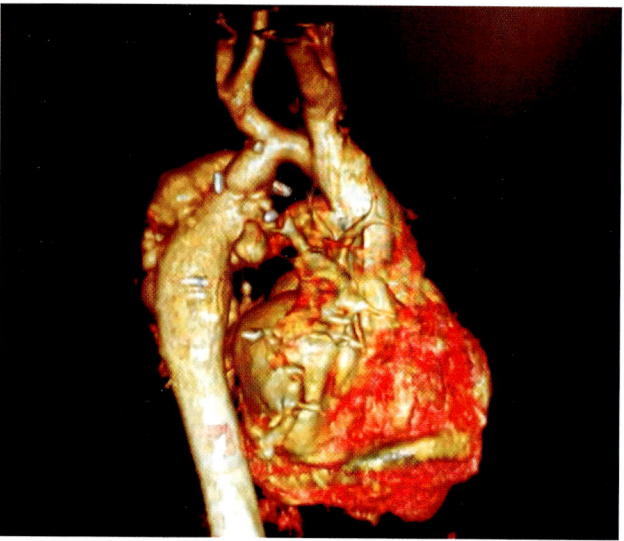

Fig. 12.107: CT volume rendering normal dimension of rest of the aorta

Radiological Diagnosis

Pseudoaneurysms in postoperative descending aorta with aortic wall thickening and irregularity likely ulcer formation, dilated LV, vegetation in descending aorta, splenic abscess, infarct and hepatic arterial tortousity in periportal region—precursor of periportal fibrosis. Overall features suggest infective etiology.

Medical Management

Patient was kept on medical management with mainly antibiotics and planned for surgical intervention after a period of one month.

Index

A

Aberrant right subclavian artery 14
Amyloid cardiomyopathy 51
Aortic coarctation 115, 176
Aortic insufficiency 129
Aortic valve placement 176
Arrhythmias 99
Arteriovenous fistulas 122
ARVC 1
Axillary artery aneurysm 164

B

Behçet's syndrome 131

C

Cardiac myxomas 71
Cardiomegaly 177
Cardiovascular MR 93
Churg-Strauss syndrome 97
CKD 124
Coarctation 115
Collaterals 181
Confluent pulmonary branches 8
Constrictive pericarditis 24

D

D-Dimer 124
Dilated RV 125
Dissecting flap 128

E

Echocardiography 115
Embolectomy 120
Eosinophillic endocarditis 45

F

False lumen 129
FDG-PET 99
Fibrosis 184
Fontan-operation 13

G

Global hypokinesia 23
Glucose 6 PD deficiency 107

H

Hamartoma 104
HCM 37
Hypereosinophilia 46

I

IE 43
Ischemic cardiomyopathy 54

L

Large VSD 115
Left atrial mass 69
Left subclavian angioplasty 119
Left ventricular apical clot 65
Left ventricular hypertrophy 35
Left ventricular thrombi 58
Lipoma of the thymus 104
Lipothymoma 104
Löffler syndrome 45
LV qualitative analysis 125
LV true aneurysm 81
LVAD 85
LVOT 89

M

MAPCAs 14
Mitral valve replacement 38
MRI tagging 89
Mycotic aneurysm 141
Myocardial infarction 81
Myocardial tagging 91
Myocarditis 54

N

Non-interrupted IVC 8

P

Papillary fibroelastoma 38
Pericardial
  calcification 28
  diseases 24
  effusion 29, 129

Pericardiectomy 23
Peripheral bypass graft 121
Periportal 184
Portal vein thrombosis 47
Pressure gradient 125
Pseudoaneurysm 82
Pulmonary atresia 14
Pulmonary embolism 125
Pulmonary hypertension 124
Pulmonary stent placement 7
Pulmonary tuberculosis 26
Pulmonary valvular atresia 8
Pulmonary vasculitis 131

R

Renal amyloidosis 52
Renal cell carcinoma 77
Respiration-related ventricular septal shift 25
Rheumatoid arthritis 46

S

Sarcoidosis 54

Septal bouncing 28, 30
Septal shift 25
Sternal dehiscence 101

T

Tagging 30
Takayasu arteritis 155
Tetralogy of Fallot 5
Thrombi 46
Thrombophlebitis 130
Thymolipoma 104
Thymolipomatous 104
TOF repair 8
Trace tricuspid regurgitation 18
Transmural enhancement 87
Type-A dissection 128

V

Vegetation 184
Velocity encoding study 125
Ventricular thrombi 63